THE BONEHUNTERS' REVENGE

BOOKS BY DAVID RAINS WALLACE

The Dark Range: A Naturalist's Night Notebook

Idle Weeds: The Life of a Sandstone Ridge

The Klamath Knot: Explorations in Myth and Evolution

The Wilder Shore (with Morley Baer)

The Turquoise Dragon: A Novel

The Untamed Garden and Other Personal Essays

Life in the Balance: A Companion to the Audubon Television Specials

Drylands (with Philip Hyde)

Bulow Hammock: Mind in a Forest

The Vermilion Parrot: A Novel

The Quetzal and the Macaw: The Story of Costa Rica's National Parks

Adventuring in Central America

The Monkey's Bridge: Evolutionary Mysteries of Central America

The Bonehunters' Revenge: Dinosaurs, Greed, and the Greatest Scientific Feud of the Gilded Age

THE BONEHUNTERS' REVENGE

Dinosaurs, Greed, and the Greatest Scientific Feud of the Gilded Age

David Rains Wallace

HOUGHTON MIFFLIN COMPANY

Boston New York 1999

Library of Congress Cataloging-in-Publication Data
Wallace, David Rains, date.
The bonehunters' revenge: dinosaurs, greed, and the greatest
scientific feud of the gilded age / David Rains Wallace.
p. cm.
Includes bibliographical references and index.
ISBN 0-395-85089-4
1. Cope, E. D. (Edward Drinker), 1840–1897. 2. Marsh, Othniel
Charles, 1831–1899. 3. Paleontologists—United States Biography.
4. Fossils—West (U.S.)—Collection and preservation—History
—19th century. I. Title.
QE707.C63W35 1999
560'.978'09034—dc21 99-31904 CIP

Printed in the United States of America

Book design by Julia Sedykh

QUM 10 9 8 7 6 5 4 3 2 1

To the many who are unable to believe that man was especially created in the image of God . . . in a society that has lost its simple faith in progress but nevertheless remains committed to the belief that "something ought to be done" about all our problems, and can be . . . the tragic spirit can promote a saving irony, in the perception of the naive or absurd aspects of this belief; a spirit of compassion through the knowledge of irremediable evils and insoluble dilemmas; and a spirit of reverence, for the idealism that keeps seeking truth, goodness, and beauty, even though human ideas are not everlasting.

— HERBERT J. MULLER,
The Spirit of Tragedy

How long, how long, in infinite pursuit
Of this and that endeavor and dispute?

— EDWARD FITZGERALD,
The Rubaiyat of Omar Khayyam

Acknowledgments

A number of scientific and conservation professionals gave generous help on this project. Greg McDonald, unit paleontologist at Hagerman Fossil Beds National Monument, helped put me in touch with other paleontologists and let me gain some rudimentary experience of excavation work at the monument's fascinating and scenic "horse quarry." Spencer Lucas, curator of geology and paleontology at the New Mexico Museum of Natural History, spent a day showing me the San Juan badlands and the museum's extraordinary early Tertiary fossil collections. Brent Breithaupt, curator of the University of Wyoming Geology Museum, made my visit to Como Bluff and Medicine Bow a highly informative experience. I'm particularly grateful to Greg, Spencer, and Brent for taking the time to review and comment on parts of the manuscript and to provide me with advice and help on references.

Rachel Benton, paleontologist at Badlands National Park, took time off to answer my questions during an impromptu visit, as did Ruthann Knudson, unit manager at Agate Fossil Beds National Monument in New Mexico. Mark Hertig, paleontologist at Agate Fossil Beds, was extremely helpful in suggesting leads during a phone conversation. Martin Lammers, historian at Fort Bridger State Historic Site, dropped his own excavations when I suddenly appeared so that he could help me go through the site's microfilm and other materials in search of evidence. Ulrich's Fossil Quarry at Fossil Station, Wyoming, gave me a replacement *Priscacara* fossil when the one I'd been working on broke. Harry Green, professor of herpetology at the University of California at Berkeley, lent me useful references on E. D. Cope's career.

I'm particularly grateful to Alan Mann, professor of anthropology at the University of Pennsylvania and curator of the University Museum of Anthropology, for his generous help with the vexing question of Cope's remains. He answered in detail my numerous, and sometimes ill-informed, questions. After doing some research on his own, he reviewed part of my manuscript. Ted Daeschler, curator of paleontology at the Philadelphia Academy of Sciences, also was helpful on this matter, as were Mark Frazier Lloyd, director of the University of Pennsylvania Archives and Records Center, Tom Huntington, editor of *American History,* and Earle E. Spamer, editor of *Proceedings of the Academy of Natural Sciences of Philadelphia.*

Judith Ann Schiff, chief research archivist at the Yale University Library, provided me with useful research materials. Mary Ann Turner, collection manager, and William R. Massa, Jr., public services archivist, were also most helpful. At the Peabody Museum, John H. Ostrom kindly and promptly responded to the question I addressed to him.

I owe thanks to Barbara Mathe, senior special collections librarian at the American Museum of Natural History, and to Ruth Sternfeld, curator of the Osborn Library at the American Museum, for kindly directing me to some previously unpublished Ballou-Osborn correspondence.

I also wish to thank Carol M. Spawn, manuscript and archives librarian at the Philadelphia Academy of Sciences, for responding to my inquiries.

This book would have been impossible without the work of many writers who have published on the Cope-Marsh feud, particularly Henry Fairfield Osborn and Charles Schuchert and Clara LeVene, who wrote biographies. Jane Pierce Davidson's extensive research for her recent Cope biography also turned up a great deal of helpful material. I'm grateful to Url Lanham, Robert Plate, and the late Wallace Stegner, whose books first sparked my interest in the subject, and to Elizabeth Noble Shor, whose transcription of the *New York Herald* articles saved me a great deal of time and eyestrain. The authors of the many other books I've consulted are too numerous to mention here.

This book also would have been impossible without the services of the University of California at Berkeley, particularly the general collections and periodical room of the Doe Library, as well as the Bancroft, Anthropology, Biosciences, Ethnic Studies, and Earth Sciences libraries. The New York Public Library, the Berkeley Public Library, the library of Ohio State University at Columbus, the Stanford University Library, and the library of Oregon State University at Eugene also were excellent sources.

Finally, I wish to express gratitude to my agent, Sandy Taylor, and my editor at Houghton Mifflin, Harry Foster, for their interest, friendliness, support, and advice, which were the most important contributions to the creation of this book.

For any misinterpretations, misquotations, factual errors, or other flaws, of course, I take full responsibility.

Contents

Illustrations

Page 108: *Marsh met with the Oglala chief Red Cloud in New Haven in 1883. Red Cloud presented him with a peace pipe as a token of esteem.*

Courtesy of Manuscripts and Archives, Yale University Library

Page 137: *The British evolutionist T. H. Huxley drew this sketch of a primeval man riding a primeval horse when he visited Marsh's lab in 1876.*

From Charles Schuchert, *O. C. Marsh: Pioneer in Paleontology*

Page 159: *A watercolor of the Como Bluff sauropod dig by Arthur Lakes.*

Courtesy of the Peabody Museum of Natural History, Yale University

Page 160–61: *Marsh reconstructed his famous genus,* Brontosaurus, *from the Como bones, a mistake paleontologists overlooked for a century.*

Courtesy of the Peabody Museum of Natural History, Yale University

Page 173: *John Wesley Powell, the U.S. Geological Survey director, wanted the government to help Indians and control Western settlement—a tall order for the Gilded Age.*

U.S. Department of the Interior, Geological Survey

Page 211: *William Hosea Ballou in 1910. A freelance hack, Ballou invented a distinguished journalistic and scientific career for himself.*

LuEsther T. Mertz Library, New York Botanical Garden

Page 259: *This 1890* Punch *cartoon of "Ringmaster Marsh" pleased its subject, but being caricatured as a mountebank had its disadvantages.*

Punch, September 13, 1890

Page 274: *Charles Knight's 1897* Century Magazine *illustration of dinosaurs fighting was probably Cope's idea.*

Courtesy of the Department of Library Services, American Museum of Natural History

Page 291: *The essayist Loren Eiseley took Cope's skeleton into his University of Pennsylvania office for "safekeeping," but may have lost part of it.*

Courtesy of Henry Groskinsky

THE BONEHUNTERS' REVENGE

Prologue: Assassination by Newspaper

In January 1890, the *New York Herald* devoted a great deal of space to a squabble between two paleontologists. Professor E. D. Cope accused Professor O. C. Marsh of scandalous crimes against science and morality, and Marsh responded by accusing Cope of worse. A crowd of other scientists and officials jumped or were dragged into the fracas, trading accusations, threats, and insults through column after close-printed column for three weeks, to the bemusement of readers who were, at most, vaguely aware that a paleontologist was a kind of hybrid biologist-geologist who studied evidence of prehistoric life.

Occurring years before dinosaurs became good copy, this "fossil feud" was not the usual stuff of 1890s mainstream journalism. The readers of some academic review might have expected it (Cope and Marsh had been fighting in such reviews for years), but the *New York Herald* had been the nation's leading newspaper for decades; its circulation only recently had become challenged by that of Joseph Pulitzer's *World*. In what was regarded as *the* metropolitan daily, a month of paleontological squabbling must have seemed a bizarre lapse, a kind of journalistic *petit mal*, and the affair, predictably, was little noticed by a public better versed in the natural history of love nests than of extinct beasts. Its descent into the oblivion of most newspaper scandals also must have seemed predictable; it was certainly craved by the officials and academicians who'd become caught in it. A 1902 book entitled *Leading American Men of Science* contained biographies of both antagonists, but didn't mention the feud. "It was an embarrassment," wrote the historian James Penick, "and held to be unrelated to the achievements of either man."

Yet the "bone war" did not lapse into oblivion. The feuding paleontologists, Cope and Marsh, had been the leaders in their field, and, as time passed, their rivalry became legendary as a colorful if cautionary sideshow to scientific history. "The most important feud . . . hindered and hampered the younger generation for years," commented William Berryman Scott, professor of geology and paleontology at Princeton from 1884 to 1930, and himself an unhappy figure in the *Herald* affair. "Even yet, its effects persist, although in no very important ways, and crop out when one is least expecting them." The legend was mainly oral at first, but it crept into print after Cope's friend, the influential paleontologist Henry Fairfield Osborn, published the lengthy biography *Cope: Master Naturalist,* in 1931. The book was meant to rescue Cope from the disrepute partly resulting from his role in the *Herald* squabble, but it prodded a Marsh ally, Charles Schuchert, professor of paleontology at Yale, to even the score by publishing, in 1940, an equally lengthy biography, *O. C. Marsh: Pioneer in Paleontology.*

Since then, an unsteady but persistent stream of publications has established the feud as a kind of hairline crack in American science, suggestive of obscure stresses within. *The New Yorker* recounted it in a 1962 "Onward and Upward with Science" piece, and several popular books about it appeared in the next two decades. In the 1980s and 1990s, books and articles on new dinosaur theories mentioned it many times. Jane Pierce Davidson's Cope biography, *The Bone Sharp,* came out in 1997, and I often heard of writers and scientists working on new Cope or Marsh books while I did the research for this one. A 1998 book included the rivalry among science's "ten liveliest debates ever," along with Galileo's struggle with

Pope Urban VIII and Newton's arguments with Leibniz. The feud has persisted in oral tradition as well, and its story has taken on strange forms, as stories do in the amorphous realm of hearsay. As the feud has receded in time, hearsay has mingled with history, adding a distinctly mythic element that consistently colors the published versions.

The bone war crept into my own consciousness in much the same way that it entered history. I kept coming across it while reading about paleontology, and it gradually evolved from a diversion to a fascination. The fact that the two paleontologists discovered and named many of the creatures that first interested me in the prehistoric past—not only dinosaurs but the giant marine reptiles that shared the Mesozoic Era with them and the strange mammals of the subsequent Cenozoic Era—contributed to the fascination. So did the circumstance that they made most of their discoveries in the West during the Indian wars. Yet many other paleontologists have discovered strange creatures in adventuresome places. Something deeper than scientific and historical picturesqueness underlies the abiding interest in Cope and Marsh: competition. Their rivalry was virtually unrestrained, without the checks and balances that cultures usually put on such disagreements, and in this they were typical of late nineteenth-century America despite their esoteric profession. In competing for a natural treasure—the abundant, unknown fossils of the western badlands—they might have been timber barons or mining tycoons, and their story is an example of a primal instinct that underlies the highest human endeavor along with the lowest.

Cope and Marsh are likely to fascinate anyone who has had to confront the myriad facets of professional jealousy—its sinking sense of annihilation, sterile brooding, impotent anger. It is an emotional abyss that is alluring as well as disturbing, and there is a temptation to jump in, to engage in intrigue, publish attacks on real or imagined rivals, and so forth. Like most people, I lack the willpower for a wholehearted plunge, but I've wondered what it would be like to let full-blown vanity sail me out over the steep canyons of unbridled rivalry. Drawing back seems as ignoble as letting go, since it doesn't end the base emotions; it merely manages them. In Cope and Marsh, I saw a classic story of those who had not drawn back, and I wanted to explore it. This was partly as ignoble as my own recoil from the abyss; I wanted to enjoy vicariously the vanity balloon ascension while witnessing the historical punishment of the daredevils who had risked it. But there may have been something of the nobler instinct that draws people to tragedy. I wanted to know what, if any, wisdom or transcendence might lie *beyond* the folly of unrestrained rivalry.

The 1890 *Herald* evidently was on to something more than an esoteric argument, which raises a question of why the paper devoted so much space to the feud. Memoirs and histories largely have attributed the *Herald*'s role to the only bylined author of the coverage, an associate of Cope's named William Hosea Ballou. Having assumed that Ballou was on the *Herald* staff, they took for granted the newspaper's participation. Evidence, however, suggests that Ballou was not a *Herald* staffer, that he was not even a professional journalist; he was a barely competent free-lancer who wrote only part of the *Herald*'s bone-war coverage and bungled that, faking interviews and infuriating named sources. An eminent paleontologist, Alfred S. Romer, later said of Ballou that if scientists "merely passed the time of day with him, they were liable to be misquoted for a column or so." Since it seemed unlikely that this hack had controlled the editorial policy of a leading daily, I began to suspect that a much larger figure lurked below Ballou's minnow-like one, a man as monstrous in his way as the paleontologists' plesiosaurs and mosasaurs. Like many monsters, this one preferred to operate in the background, and I found little uncircumstantial evidence of his involvement. But as I read what is known about him, I sensed that he offered the beginning of an explanation not only of the *Herald*'s involvement in the feud, but of its historical resonance.

James Gordon Bennett, Jr., the *Herald*'s owner and publisher from 1868 to 1918, may be the most underestimated American figure of the late nineteenth century. More than any journalist before

William Randolph Hearst, who imitated him, Bennett not only reported history but *made* it, and the period from 1870 to 1900 bore his unmistakable stamp. History would have been quite different, for example, if Bennett hadn't sent Henry M. Stanley to Africa. Yet Bennett's character was so repellent that it has obscured his historical role. His only recent biographer, Richard O'Connor, introduced his book *The Scandalous Mr. Bennett* by reflecting that the protagonist deserved a serious scholarly study but wasn't going to get one from *him* because he was too preposterous.

Instead, O'Connor luxuriated in what might be a case study from Krafft-Ebing's *Psychopathia Sexualis.* "On midnight rides, careering down country roads, he would often strip off his clothing," he wrote of the youthful Bennett. "Stark-naked in the box, he would drive his four-in-hand along the turnpikes, cracking his whip and yelling his head off in delirious pleasure." According to O'Connor, Bennett had an owl fetish; he had the birds tattooed on his mistress's knees, and planned to be buried in a 125-foot, granite owl statue with "quite ferociously" glowing eyes. Bennett's photos support O'Connor's Krafft-Ebing vision, showing a kind of debauched dragoon with a strangely narrow head punctuated by a fleshy nose and small, close-set, furiously staring eyes. In fact, Bennett was an alcoholic sociopath who committed a crowning sexual peccadillo so embarrassing that it still makes a reader squirm, over a century later. On New Year's Day of 1877, he drunkenly urinated in the fireplace (or grand piano) of his fiancée's parlor during a gala party. For this, he was horsewhipped by the fiancée's brother and permanently banished from New York society.

So it's no wonder historians have shied away from James Gordon Bennett, Jr. As it happens, he was a *controlled* alcoholic sociopath who knew what he wanted and pursued his desires with the sociopath's singleminded energy. He may have deliberately committed parlor urination to escape his engagement without incurring a costly breach-of-promise suit. He also was a fabulously rich alcoholic sociopath who could enact his desires on a grand scale. He had virtually absolute control over his newspapers.

Bennett "regarded every word printed by his papers as emanating from himself," O'Connor wrote. "If someone else happened to put those words on paper, they came from what he was pleased to call a 'hired brain.' His employees simply had to get used to the idea of being treated as so many intellectual zombies." A *Herald* alumnus, Albert Stevens Crockett, recalled that "each man who worked on the *Herald,* no matter what his office or how important his job, must perforce work to please Bennett, and the public only through him." Like other dictators, he had no problem acquiring lackeys. "There was something compelling about the tigerish proprietor, with his fickleness and brutality," wrote Don Carlos Seitz, an editor on the rival *New York World.* "Thus, for all their contemptuous ill treatment, men were attracted to the *Herald.*"

Bennett's father had founded the *Herald* in the 1830s as a pioneer of modern journalism's mélange of advertising, news, and sports as well as its emphasis on crime and other sensations. The paper first captured public attention with a story featuring a rapt description of a murdered prostitute's nude corpse as surpassing "in every respect, the Venus de Medici." Another popular innovation was the list of classified "personals" ads, which read very much like today's: "Woman finds paddling her own canoe dreary task, seeks manly pilot." Bennett Sr.'s idealistic competitor, the *Tribune*'s Horace Greeley, called him a "low-mouthed, blatant, witless, brutal scoundrel." Yet Bennett used his smelly cash cow to finance an efficient news-gathering organization. The *Herald*'s Civil War coverage was unprecedented in its speed and thoroughness.

After becoming publisher in 1868, Bennett Jr. continued his father's leadership in news-gathering as well as in profitable sensation-mongering. The *Herald* reported Custer's 1876 Little Bighorn defeat four days before the other New York papers did, and before the War Department even knew of it. Bennett Jr. also went his father one better by his willingness to *make* news as well as report it. Sometimes he simply lied, as when, in November 1874, the *Herald* reported that escaped zoo animals had mauled 249 people in Central Park. "The coming of the tiger was something terrible," the full-page article panted. "I never shall forget the awful, splendid look of him as he landed with a spring in the thick of them" (Other "wild carnivorous beasts . . . still at large" included a manatee, an opossum, and "the paisano, a vicious beast, said to be on the west side of town," but many readers panicked anyway.) More often, Bennett chose a real sensation and injected one of his zombies into it as actor-reporter.

The most famous zombie, of course, was Stanley, a nonalcoholic sociopath who might have ended on the gallows (his crimes included extortion, assault, and desertion from both the Confederate Army and Union Navy) if the *Herald* had not hired him in the late 1860s. His epic 1871 search for the missionary-explorer Dr. David Livingstone was more a product of Bennett's imagination than of Stanley's, as the publisher was quick to insist. Stanley had cared little about Livingstone until Bennett summoned him to his Paris hotel and declared, "I think he is alive and that he can be found, and I am going to send you to find him." And Stanley's expedition was only one of many Bennett financed to the ends of the earth—central Asia, Siberia, the North Pole. The publisher's obsession with penetrating the unknown not only reflected the nineteenth century's exploration vogue, but magnified it. "In recognizing this new curiosity, which eventually burgeoned into all sorts of national ventures overseas, militant and peaceful alike," wrote Richard O'Connor, "Bennett demonstrated the grasp—if not the temperament—of a great editor."

Whether or not Bennett was a great editor, it's hard to think of another who had a similar effect. Perhaps one reason he hasn't been called great is that, unlike Stanley, he made little pretense of acting for the benefit of civilization. He said he supported exploration to sell newspapers, but he was disingenuous even in that cynical attitude. Not a businessman at heart, he milked the *Herald* of $30 million and left it nearly bankrupt at his death. At heart, Bennett seems to have been not merely indifferent, but hostile, to civilization's benefit. His zoo hoax suggests a fascination not with taming savagery but with savagery's potential for dissolving civilized constraints, which he found irksome. Livingstone's saintly aura probably interested him less than the allegations of brutality and consorting with native women that led Joseph Conrad to model, in part, his *Heart of Darkness* anti-hero, Kurtz, on the missionary.

Bennett might have found some satisfaction by exploring Africa himself, Kurtz-fashion. He didn't lack physical courage; he raced his yacht across the Atlantic more than once. But he was unwilling to forgo luxuries, and knew he lacked the stomach for the extreme hardships of nineteenth-century expeditions. This didn't stop him from bitterly envying the fame of those who did explore, and from feeling that he deserved it. "He could only sponsor them," O'Connor wrote, "and sulk in the shadow of their renown." He turned against Stanley quickly, which was evident in his publishing a scathing review of the explorer's 1872 New York lecture appearance, and flying into a rage when he read of Stanley's exploits in other papers. In 1891, he sent a reporter to the Tyrolean resort where Stanley was vacationing in an attempt to confirm rumors that the explorer beat his wife. ("My God," said Stanley when he realized what the reporter was after, "I used to do that.")

Bennett's wealth and craziness, interacting with his boundless envy, made him behave like a trickster god, a lord of misrule. Not even innocent bystanders escaped his spite. In swanky restaurants, he would amuse himself by yanking off other diners' tablecloths while well-bribed waiters stood by. Esoteric as it seemed, the two paleontologists' squabble was just the thing to attract his malice. Its scientific antagonists were the kind of men he most envied, highly respected explorers who had helped to open a new world to civilization. That world, the fossil-rich plains of western North America, may have been less sensational than Africa, but it was more significant. It not only contained vast new geographical spaces; it embraced vaster temporal ones, the hundreds of millions of years in which monsters, not men, had ruled the earth.

If the paleontologists lacked Stanley's huge celebrity, one of them, Professor O. C. Marsh of Yale, had attained a statesmanlike eminence as an adviser to presidents, protégé of powerful congressmen and bureaucrats, and confidant of scientific mandarins like Thomas Henry Huxley. Indeed, Marsh often had figured prominently in the *Herald*'s pages, a sure way of arousing its owner's jealousy. Bennett would have been quick to perceive that the bone war was *about* Marsh's eminence. The rival, E. D. Cope, envied Marsh as bitterly as Bennett envied Stanley, and felt that Marsh had attained a distinction that rightly belonged to him—and had done so at his expense. For two decades, Marsh, in fact, had been using his political power to block Cope's scientific aspirations. Now Cope was trying to use the hack Ballou, whom he naïvely characterized as "a rough customer," to wreak revenge on Marsh by destroying *his* scientific reputation. It was the scientific equivalent of a palace assassination, and if Marsh and Cope had been Hellenic or Renaissance princelings, their quarrel might have attracted a Euripides or Shakespeare. Bennett's barbaric figure looms behind them like Dionysus in *The Bacchae,* a *deus ex machina* inciting antagonists to greater hubris.

The newspaper spectacle of professorial beard-pulling may have been comic, but the bone war was tragic in that it poisoned the lives of both men, neither of whom survived the *Herald* scandal by a decade. The paleontologists' quarrel also was like classic tragedy in enacting social as well as personal conflicts. Bennett was the embodiment of the conflicts, a *boulevardier* who marketed fantasies of ladies devoured by tigers in what his paper called "a shocking Sabbath carnival of death."

America has played out Bennett's—and Europe's—conflicted desire to escape civilization while simultaneously imposing it on "dark continents." Nineteenth-century United States history is a story of colonies that freed themselves from a transoceanic European empire only to fall subject—through competition among themselves—to a transcontinental American one. Originating from states so different that they might have had separate nationalities, Cope and Marsh embodied this competition. A Pennsylvania Quaker, Cope rose from a tradition of "colonialist" tendencies toward rural, anarchic identification with the American continent. A native of residually Puritan Massachusetts, Marsh came to represent an opposing "imperialist" tendency toward an urban, hierarchical subordination of such "nativism." In a way, the bone war was an intellectual variant on the Civil War that just preceded it, one largely enacted in that other great arena of North American conflict, the West.

Indeed, the course of American empire might have been different without the feud, because Marsh's and Cope's fossil-hunting played a significant role in the geological surveys that prepared the West for conquest. As Wallace Stegner showed in his classic study of the period, *Beyond the Hundredth Meridian,* the *Herald*'s trumped-up paleontological scandal contributed to the failure of a plan for scientifically limited settlement advocated by John Wesley Powell, director of the U.S. Geological Survey. Powell was allied with Marsh, and the newspaper squabble damaged the reputations of both, contributing to the downfall of each. "It is doubtful that any modern controversy among men of learning has generated more venom than this one did," wrote Stegner.

Although Powell's plan was a product of federal bureaucracy, it might have led to settlement more in keeping with the anarchic, agrarian tradition of colonial America than the centralized, industrialized system that came to dominate the West after droughts defeated the *laissez-faire* boosterism of Powell's opponents. At least, his policies might have proved more environmentally sustainable than the ones that haphazardly emerged. So the *Herald*'s 1890 lunge into paleontological backstabbing helped to ensure that the issue of North America's settlement would remain "unsettled" to this day, as rural anarchism still wars with urban hierarchy in land politics. Like the Civil and Indian wars, the bone war lingers in a United States of ethnic-regional militias and antifederal terrorism.

1

Prodigy and Heir

When Edward Drinker Cope was born into a well-to-do Philadelphia Quaker family in 1840, he inherited a century-old scientific tradition. George Fox, the Society of Friends' seventeenth-century founder, had encouraged his followers to garden and to study plants as a way of knowing God through His creation. Barred from universities and official posts for their nonconformism, the temperate, industrious Quakers prospered as merchants and physicians, and used trade networks to promote the collection and study not only of plants but of all natural phenomena. One of the most important connections was the one linking England with William Penn's North American colony. In the mid-eighteenth century, a London merchant named Peter Collinson dominated this informal structure, and asked his Pennsylvanian co-religionists to supply specimens and information that he could introduce into his scientific circle. Collinson was a corespondent of Linnaeus, the century's major botanist, so that circle was very high.

Collinson's main source of American information and specimens was a down-to-earth Quaker farmer named John Bartram, who, it was said, became interested in botany when, out plowing, he had noticed the complexity of a daisy. With Collinson as middleman, buying an endless stream of plants, shells, live animals, and other "natural productions," and passing them on to noble patrons, Bartram became a model New World naturalist, full of frontier self-reliance and ingenuity and of devout optimism. In the Friends' tradition, he saw "the immediate finger of God" in nature, and believed that "a portion of universal intellect diffused in all life." His botanizing was a way both to satisfy his intellectual curiosity and to praise "the living God, the great I am." Scientific and religious enthusiasm made a robust mixture that sustained him through dozens of expeditions, often alone, beyond the narrow coastal band of European settlement. Undaunted by Indians, although they had killed his father, Bartram, between the 1730s and 1760s, explored upstate New York, the Ohio River, the Carolinas, Georgia, and Florida.

As Bartram's skill and reputation grew, so did his eminence in the colonial intellectual establishment, which, if it did not always share his Quakerism (Bartram himself broke with his meeting by denying Christ's divinity), certainly shared his deist perception of God in nature. He was a close friend of Benjamin Franklin, and helped him found the American Philosophical Society in 1743. They made Philadelphia the center of colonial science, even though neither had had professional training in any scientific discipline, training that was largely unavailable at the time and considered unnecessary. With an almost unknown world before him, the colonial naturalist was free to study anything without seeking authority. Bartram's writings described not only plants but fauna, soils, rock formations, ethnology, meteorology, medicines, and whatever else interested him.

The historian Daniel Boorstin saw this as a significant break from the Old World's more elite, institutionalized scientific community. "The ideal of knowledge which comes from natural history was admirably suited to a mobile society," Boorstin wrote in *The Americans: The Colonial Experience:* "Its paths did not run only through the academy, the monastery, or the university; they opened everywhere and to every man." Indeed, Boorstin regarded direct observation of nature—as opposed to erudite hypotheses—as distinctively American. "American life quickly proved uncongenial to any special class of 'knowers,' he wrote. "Men were more interested in the elaboration of experience than in the elaboration of 'truth'; the novelties of the New World led them to suspect that elaborate verification might itself mislead . . . No American invention has influenced the world so powerfully as the concept of knowledge which sprang from the American experience." Josephine Herbst, Bartram's biographer, called colonial American naturalists "whole men confronting a whole world, not human beings floating in a culture medium."

Philadelphia's scientific pre-eminence continued after the American Revolution. John's son William also had explored the Southeast, and his 1789 book about it, *Travels,* became the first major text of the United States' naturalist tradition ("a future *Biblical* article," as Thomas Carlyle put it). Perhaps influenced by Rousseau, William went beyond his father and Franklin. He was the first colonial writer to integrate Native Americans with the deist vision of God in Nature, as seen in his sympathetic picture of Florida Creek culture. *Travels* evokes a naked Seminole who "reclines and reposes under the odiferous shades of Zanthoxylon, his verdant couch guarded by the Deity, Liberty and the Muses inspiring him to wisdom and valor." Such evocations did much to shape the Romantics' enthusiasm for untamed nature and "primitive" man. Wordsworth, Coleridge, and Chateaubriand borrowed heavily from *Travels.*

The Bartrams' botanical garden outside Philadelphia fostered a new generation of naturalists in the United States. Referring to their activities, Thomas Jefferson, after serving as a delegate to the 1787 Constitutional Convention, wrote to his friend William: "I long to be free for pursuits of this kind instead of the detestable ones in which I am now laboring." Jefferson's preference for natural history over statecraft was hardly a pose; his *Notes on the State of Virginia* is another early classic in the field. As he rose to national leadership, natural history figured largely in his ideal of the United States as a society of enlightened farmers, a "nature's nation," wherein an educated, genteel population would not become urbanized and etiolated, but would maintain a self-sufficient agrarian base. When he was President, he sent Meriwether Lewis to study with William Bartram and other Philadelphia naturalists before setting out to explore the Louisiana Purchase. And after leaving office, he shaped his University of Virginia, the first nonecclesiastical college in the Americas, very much in the Franklin-Bartram mode. It was to be an institution of self-guided learning, without matriculation or degrees, without formal hierarchy.

In 1812, William and his friends founded another major naturalists' association, the Philadelphia Academy of Natural Sciences. At the same time, Philadelphia saw the first successful attempt to diffuse natural history to the public. The local artist Charles Willson Peale had painted famous portraits of Washington and other Revolutionary heroes, but when he was asked to draw mastodon fossils from Kentucky, he became so interested in them that he gave up portraiture in favor of establishing America's first public natural history museum. Installed in the room where the Declaration of Independence had been signed, it displayed fossils and realistically posed animals in lifelike settings (although it also included a five-legged cow, costumed monkeys, and the occasional song and dance act). Optimistic about its role, he told Jefferson, who tried and failed to get government sponsorship for it, that "such a museum, easy of access, must tend to make all classes of people in some degree learned in the science of nature without even the trouble of study . . . Nature, opening new ranges of beauty and understanding, length of days and joy of life." In 1810, a visitor named Catherine Fritsch fulfilled Peale's hopes when she gushed: "Here we could observe abundant instance of the wisdom of God in his Creation, as we viewed, with astonishment, the many different animals, birds, and fish, and the infinite variety of exquisite butterflies and insects."

Edward Cope's family seems not to have had a close social connection with Philadelphia scientific circles, although his father, Alfred, belonged to both the Philosophical Society and the Philadelphia Academy. Edward grew up, however, in an atmosphere infused with the Bartrams' spirit. His father presided over an eight-acre estate, Fairfield, just outside the town, and there he botanized and, according to Henry Fairfield Osborn's *Master Naturalist,* lent his children "a splendid collection of books within the house, and his own active intelligence out of doors." Descended from a 1686 English immigrant, Alfred's father, Thomas, had so prospered in freight shipping that his son was free to devote himself to such philanthropic pursuits as educating Indians and financing a Philadelphia zoo.

Edward quickly began to behave like an heir apparent to the Bartram tradition. At six, he started recording his natural history observations in journals, letters, and drawings. One of the first letters concerned an enthralling visit to the Peale Museum. He wrote of the visit to his grandmother in 1846, "I saw Mammoth and Hydrarchus. Does thee know what that is? It is a great skeleton of a serpent. It was so long that it had to be put in three rooms. There was a stuffed crocodile, and an alligator, and the crocodile looked the ugliest and fiercest, and his mouth wide open. And I saw a monkey's blacksmith shop, and one had a newspaper reading the Foreign news." Cope may have been Victorian America's closest thing to a naturalist child prodigy, since the type tends to be a late bloomer (with Darwin the classic example). A playmate recalled "an incessant activity mind and body . . . People's attention was instantly caught by his quick and ingenious thought, expressed in a bright and merry way. His mind reached in every direction for knowledge."

Edward's later development confirmed his early promise. A year after his charming but childish museum letter, he wrote and illustrated a little hand-sewn journal of a boat trip to Boston with his father, punctiliously dating each entry and describing his observations with concision. "We saw some Bonetas swimming along-side the vessel," read his fifth-day entry; "they're long slim fish and twist about like eels. We saw a Man-of-war, which looked like a large jelly-fish only he was dead. The night before we saw a great many lights in the water which were made by Jelly-fish and here is a picture of it." By 1849, when he went to a Philadelphia Quaker day school, Cope had become a habitué of the Philadelphia Academy's scientific museum, which he described that year as though working up a curatorial inventory: ". . . several small skulls of birds of different sizes and forms; some of them had red, black, and white bills. Among them were about five skulls of toucans." More important, he had begun to bring to the museum for identification animals he'd collected in the countryside.

After his father sent him to the Friends' Boarding School at Westtown in 1853, Edward wrote to his sisters to complain of the natural history facilities: "I had expected a handsome large room, but instead of that it was an old-fashioned room with whitewashed walls and ceiling." And apparently he knew more about reptiles than did his instructors. In another letter, home, he offhandedly mentioned a scientific name for a turtle which a "Master Davis" had been unable to identify. In a later letter, to a cousin, he described a well-informed outing for a teenager: "I traced the stream for a considerable distance upon the rocky hillside, my admiration never ceasing, but I finally turned off into the woods toward some towering rocks. Here I actually got to searching for salamanders and was rewarded by capturing two specimens of species which I never saw before alive. The first (*Spelerpes longicauda*) is a great rarity here. I am doubtful of it having been previously noted in Chester County." He would publish his first professional paper, "On the Primary Divisions of the Salamandridae with Descriptions of New Species," in the *Proceedings of the Philadelphia Academy of Sciences* when he was nineteen. According to Henry Fairfield Osborn, it "instituted important modifications in the accepted classification of Salamders." Still in his teens, he also began working as an assistant curator at the academy.

Cope probably would have been even more precocious if his father hadn't spent the 1850s trying to rein him in. A devout Friend, Alfred "rigidly supervised" his children's conduct and religious training. Edward's letters home from Westtown were full of convoluted protestations in defense of what Alfred considered unacceptably willful behavior. "I only wish he would be a scholar with me for a month or so," he wrote to his sisters in 1856, "and could come to particulars and see what is enough to give one such conduct—I don't think he would consider me such a wicked boy. Laughing a little too loud and a great many little things go together, and make a bad conduct number." The father's inflexibility would cause Edward mental turmoil in later life, although maternal buffers ameliorated it somewhat. From the time his mother, Hannah Edge Cope, died, in 1843, Aunt Jane had cared for Edward and his sisters until Alfred married again, in 1851. Rebeccah Biddle, the stepmother, evidently was sympathetic to Edward. "She was his refuge and court of appeal when his conduct displeased his father," Osborn wrote, "and she seems to have interceded for him mildly but to good effect."

Despite his rigidity as a parent, Alfred evidently was an enthusiastic Jeffersonian, since he tried to mold his son into an educated gentleman farmer like the third President. Beginning in 1854, he sent Edward to work on relatives' farms every summer, and after Edward finished Westtown, in 1856, Alfred set him to learning agriculture full time, with the intention of buying him a farm when he reached manhood. Edward underwent this laborious education cheerfully, and seems to have enjoyed country life, but he had little interest in farming. "I have been hoe-harrowing corn some lately," he wrote to his father in 1859, "and the thought occurred to me several times as I walked slowly back and forward across the field—how much more money could a man make by applying himself to some other business or rather while engaged in some other business, during the days and weeks that the farmer pokes backward and forward across his field, earning nothing beyond the cutting of a few weeds." The next year, he asked Alfred for permission to attend lectures in comparative anatomy given by Dr. Joseph Leidy, a zoologist and paleontologist at the University of Pennsylvania. "The whole ground is gone over in winter," he wrote reassuringly, "and the knowledge of human and comparative anatomy would be of immense service to one desiring a knowledge of the proper manner of treating stock and of general comparative anatomy." Stock-raising, however, was not what he had in mind; he was soon spending all his time on natural history.

Alfred Cope—a photograph shows him as a slight, subdued man—evidently despaired of controlling his son after that year. From Edward's photos, it's not hard to see why. Even in a stiff plate from the 1870s, the younger Cope's bristling hair and sturdy body manifest extraordinary will and energy. The protuberant eyes suggest an overactive thyroid, a hint supported by his lifelong nervous crises. Alfred did buy Edward a farm, the oddly named McShag's Pinnacle in Chester County, but seems not to have protested when his son chose to put a tenant there. After Edward spent the next few years doggedly pursuing natural history, Alfred paid for him to study in Europe, although this may have been an expedient for separating the high-spirited young man from an undesirable romantic attachment and from the equally undesirable Civil War.

The scientific education that Cope acquired between 1860 and 1864 had as much of the eighteenth century about it as of the nineteenth. Without formal structure, it essentially consisted of Edward's studying collections—first at the Philadelphia Academy and Smithsonian, then at the British Museum and other European institutions—and talking with their curators. These were activities suitable to a young Enlightenment patrician with serious tastes, and they proferred no degree or other professional credentials. Cope's education was equally old-fashioned in its lack of specialization. Although he gravitated toward vertebrates, he seems to have studied living and fossil organisms pretty much as the spirit moved him. Nor did he or his father see any disadvantage in this quaint program. Opportunities for university training in natural history were still limited; as Osborn observed in the 1930s, "What great university of our day could offer the educational influences of the series of great men whose acquaintance Edward eagerly sought and whose friendship and companionship he cultivated during this transitional period?" Edward's precocious election to the Philadelphia

Edward Drinker Cope at thirty

Academy in 1861 and to the National Academy of Sciences in 1871 certainly spoke well of such influences.

Cope's genteel *Bildung* did depart from the Enlightenment model in one vital way. After the 1859 publication of Darwin's *On the Origin of Species,* the "great men whose acquaintance Edward eagerly sought" would have been less interested than their predecessors in the evidence of Divine Wisdom in nature and more inquisitive about the evidence of evolutionary change. At first, Edward apparently responded with equanimity to Darwinism, although it must have been a sharp break from his father's ideas about nature. He'd read Darwin's *Voyage of the Beagle* at Westtown in 1856, and had found it "exceedingly interesting, only it had a little too much geology in it." In Venice, at Saint Mark's Cathedral in 1863, he expressed a viewpoint that departed sharply from biblical literalism. "Near the entrance is a large flag of red conglomerate in the floor where one of the Popes received the homage of Barbarossa," he wrote to his sister. "All this seems very old, but I spied ammonite [extinct fossil cephalopods] in another piece of the same stone in the lower part of the wall. It made me grin to think how all that it had seen, from the early Doges to Marino Faliero and the white coats of Austria, were in the last few minutes of its existence."

Even more significant is that Cope seems to have grasped the profound implications of a fossil he saw in Germany that year. "The collections from Solenhofen are exceedingly interesting and numerous," he wrote to his father. "Prof. Opfel who has charge of it was the first to obtain the *Archaeopteryx lithographica,* and he believes that he has a skull which he showed me. It is very bird-like, but has a long, slender bill as *Conchiosaurus* [a dinosaur] . . . It is nearer bird than reptile but is neither." A dinosaur-like creature with feathers and wings, evidently transitional between reptiles and birds, *Archaeopteryx* was the first major fossil proof of Darwinian evolution after the publication of *The Origin of Species.*

Yet Edward also manifested considerable unease during his grand tour. "If I know myself I need every possible aid to distract myself from myself," he wrote from London in 1864, "and if I do not have it my health suffers; what it would result in if my various outlets for my activities were not to my hand I cannot tell—but I do not much doubt, in insanity." Much of this disquietude probably was a residue of his unhappy love affair, which he felt had "scorched . . . the outside" of his sensibilities "into a crust." Osborn believed that Cope shared the period's intellectual anxieties about the widening gulf between science and religion, however, and that he underwent a crisis at the tour's end. "That he had an inner consciousness of doing something wrong in his pursuit of science seems to have grown upon him," Osborn wrote, "because, just before sailing for home in the year 1864, Edward destroyed many of his scientific drawings made from various museums, and was restrained by one of his friends with difficulty from burning all his priceless European notes. Upon his return to Philadelphia he was for a time deeply religious, and made an open confession of faith at Meeting, which, according to his sister, was a great comfort to his family."

Whatever the reason, Cope sought stability back in Pennsylvania. In the fall of 1864, he became a professor of zoology at Haverford College, a position obtained through family influence, according to Jane Pierce Davidson, who noted that the college compensated for his lack of credentials simply by granting him an honorary A.M. degree in his contract. "My position here is pleasant," Edward complacently told father, "especially as I can interest considerably many of my students. Dr. Flack in Philadelphia has lent me a large lot of reptile, bird, and mammal skeletons—a foreshadowing, I hope, of what may be in the possession of the college one day." He courted and married a distant cousin in a similar sober, somewhat calculating spirit. "I have often thought it would be an advantage to me to be married in many ways, and have concluded to make a definite motion in that direction . . . ," he wrote to Alfred. "An amiable woman, not over sensitive, with considerable energy, and especially one inclined to be serious and not inclined to frivolity—the more truly Christian of course the better—seems to me to be practically the most suitable to me, though intellect and accomplishments have more charms." Luckily, his wife, Annie, proved compatible as well as "not over sensitive," so he acquired a warm home and, promptly, a beloved daughter.

As the Civil War and its difficult ethical questions ended, Cope may have felt that his real life was beginning, a life not so different from that of the Bartrams a century earlier, although more affluent and secure. With the income from his farm and his teaching, and the prospect of a comfortable inheritance, he could anticipate a career of research, travel, collecting, and correspondence, interspersed with farming and philanthropy—a Jeffersonian idyll. "I am very glad to do whatever will break the selfish nature more," he wrote about his college duties, "and fulfill the 'sell all thou hast and give to the poor.' And withal I have abundance of time to pursue such original studies as material sufficient comes to my hand."

Yet the Philadelphia and the America to which Cope returned in 1864 were very different from those in which he'd been born—even from those he had left in 1862. Slavery long had made a mockery of the Jeffersonian ideal, and the war had scorched "nature's nation" to an undeniable "crust." Philadelphia was no longer a genteel center of trade and governance, but a secondary cog in an explosive machine of railroads and factories. What was even more significant for Cope's future, the city was no longer the center of American science. The last members of the old Bartram circle, the entomologist Thomas Say (John Bartram's great-grandson) and the botanist Thomas Nuttall, had left in 1825, Say to join an Indiana utopian community, Nuttall to teach at Harvard. The very structure of natural history was changing. Say and Nuttall had been of the Bartram mold, independent explorers who roamed the Rockies and Oregon in search of new specimens. But, as the writer Joseph Kastner observed of the 1840s: "The adventurer-naturalist now was being eclipsed by the academic specialist . . . A Bartram or a Nuttall would no longer go where his dream or ambition took him, but where a professor told him to go. The romantic age of American natural history was nearly over."

By 1865, Say and Nuttall were dead, and Ivy League professors like Harvard's Asa Gray presided over rapidly coalescing academic specialties. Gray was too busy with research, cataloguing, teaching, and corresponding with colleagues like Darwin to do much exploring. "The enlightened bureaucrat had replaced the virtuoso as the benefactor—and the beneficiary—of American natural science," wrote Kastner. As the virtuoso Edward Cope wandered through the Pennsylvania woods collecting flowers, rocks, mammals, toads, fossils, and whatever else struck his fancy, he was somewhat in the position of a cultivated Southerner in 1860. Having inherited a venerable and apparently stable world, he was only dimly aware of another growing quickly to the north. Soon he would be tempted to challenge that world, however, and he would learn more about it, disagreeably.

2

Stepchild and Laggard

When Othniel Charles Marsh was born to a ne'er-do-well Yankee farmer in 1831, he was heir to little property or natural history. John Calvin, the father of what became Puritan theology, hadn't encouraged his followers to seek God in gardens, and they certainly hadn't sought Him in the American wilderness. They saw mainly His adversary there, as a Northeastern abundance of "Devil's Punchbowls" and "Satan's Knobs" still attests. Puritan divines like Cotton Mather sometimes took time off from theocratic pursuits (which, in fairness, only rarely and reluctantly included hanging the odd Quaker or other heretic) to comment on "useful natural productions," but colonial New England developed little in the way of a natural history coterie. Before 1800 Harvard and Yale laid heavy stress on theological studies.

New York had eighteenth-century naturalists like Cadwallader Colden and Samuel Mitchill, but they were isolated, in comparison with the Philadelphia group. When pioneer bird artist Alexander Wilson, another member of the Bartram circle, traveled through the South seeking subscribers to his *Ornithology* in 1808, he sold thirty-four subscriptions in Richmond alone and sixty-three in New Orleans. In the entire Northeast, he sold forty-one. "My journey through New England has rather lowered the Yankees in my esteem," wrote Wilson, a Scottish immigrant. It wasn't until Emerson and Thoreau began writing in the 1830s that the Northeast developed a distinctive literary tradition in natural history, and it owed more to European influences than to Philadelphia's scholars. Thoreau read Bartram's *Travels,* but it affected him much less than it had Wordsworth and Coleridge. Thoreau's puritanical sense that scientific knowledge somehow vitiated aesthetic appreciation would have puzzled the Bartrams.

Othniel Marsh's ancestors were typical New Englanders who just missed arriving on the *Mayflower.* The family of his father, Caleb, had reached Massachusetts in 1634; that of his mother, Mary Peabody, in 1635. Both families had stayed there, as small farmers and shopkeepers, until the early 1800s, when the Peabodys produced Mary's brother George, a financial prodigy who began as a door-to-door peddler in 1812 in Washington, D.C., and ended as a London-based banker of international influence. George, unmarried, interested himself in his relatives' affairs and helped get Massachusetts moving. He provided his sister Mary with a dowry, which made it possible for Caleb Marsh to buy a farm in Lockport, New York. Othniel was born there. Unfortunately, Caleb Marsh didn't turn out as George had hoped. Mary died in 1834, and Caleb lost heart. He returned to Massachusetts, started a shoe factory, and married again, but the business failed in 1837, and he went back to Lockport. There, he produced six more children and acquired debts that ate up most of a settlement George Peabody had set aside for Othniel and his older sister, Mary.

Othniel's stepmother had much less time for him than Edward Cope's stepmother had for him, and Othniel's youth clearly was grimmer than his Pennsylvania counterpart's. "The oldest son of a rapidly increasing household," wrote his biographer, Charles Schuchert, in collaboration with Clara LeVene, librarian of the Peabody Museum, "he was expected to be his father's mainstay in the farm work, and his reluctance to do so was a source of friction between them. He preferred to roam the fields and woods, hunting the small game then still abundant in the Lockport region." This friction, unlike that between the Copes, was never resolved. Caleb's photograph shows a standard Yankee, lantern-jawed and hatchet-faced, not very resourceful, perhaps, but not to be swayed by new-fangled notions. Like Cope, Othniel had little physical resemblance to his father. He looked more like his uncle George Peabody, with fleshy features, deep-set eyes, a high-domed brow, and a piercing expression. Caleb must have found him even harder to push around than the rocks in a field, and there's little evidence of the open communication enjoyed by Alfred and Edward Cope.

There's little evidence, in fact, of anything about Marsh's boyhood. He was not a confiding correspondent like Cope, and his occasional diaries were mainly records of "events, rather than of the writer's reaction to such," as Schuchert and LeVene put it. As the farm "mainstay," he probably attended only winter school sessions, and quit when he was sixteen, but there does seem to have been one early intellectual influence. An amateur geologist, Colonel Ezekiel Jewett, interested Othniel in the abundant fossils and minerals to be found in the banks of the Lockport Canal, although the boy was said to have been as impressed by Jewett's shooting as his geology. In 1847 Marsh went to the Collegiate Institute, in nearby Wilson, and in 1850 to the new Union School in Lockport, but distinguished himself in neither. His father called him "a boy of good natural abilities and some mechanical ingenuity—much inclined to a roving disposition."

Marsh's disposition led him away from Caleb's farm as soon as he was old enough to leave. He tried teaching at Millport, New York, but quit because he had headaches and eye trouble, and then moved back to Massachusetts to be near his sister Mary, of whom he was fond. Mary's husband, Captain Robert Waters, wrote that the twenty-one-year-old's character was "hardly formed and developed," and Othniel's diary of that year mentioned only a vague inclination to be a carpenter or surveyor. So he might have been if not for Uncle George Peabody. Some $1200 remained of his mother's marriage settlement, and when Marsh came into it, he decided to attend Phillips Academy in Andover. He spent his first year there "making little impression" and still intending to "work at the mechanical trade." Something affected him during the summer vacation, however; it may have been his sister's death during childbirth. "I changed my mind during an afternoon spent on Dracut Heights," he later wrote. "I resolved that I would return to Andover, take hold, and really study."

Othniel began collecting minerals enthusiastically, and the next year switched to the college-preparatory Latin curriculum. Considering his previous conduct, he became an achiever of surprising drive. A school friend, W. E. Park, wrote: "After Marsh really began to study, he stood first in class every term without exception. He studied intensely, but tried to make the impression that he achieved his success without any work . . . His superiority of managing practical affairs soon impressed all." Park characterized his friend's "foresight and shrewd management—shrewd with a touch of cunning in it" by recounting how Marsh had orchestrated his election to the presidency of a school society with a strategy planned a year in advance. Schuchert and LeVene observed that Othniel's 1854 diary contained little personal reflection but much "recurrence of the word success."

Marsh spent the summers of 1854 and 1855 collecting minerals in Nova Scotia, then a fashionable geological location, and also chanced upon his first vertebrate fossil, which he thought was a halibut's backbone. Later, when he was a graduate student, the fossil caused a stir because the great Harvard ichthyologist, Louis Agassiz, decided that it had characteristics of both fish and amphibians. That didn't impress Othniel. He disliked Agassiz's meddling and decided the fossil was an ichthyosaur, a common fishlike reptile, which he named *Eosaurus acadiensis.* (It later was identified as an early amphibian.) Marsh wasn't much interested in paleontology at first; mineralogy remained his passion throughout his university career, which began in the fall of 1856, when he entered Yale.

New Haven was a break from tradition. His family, including Uncle George, had gone to Harvard, but Yale's new Sheffield Scientific Graduate School attracted Othniel. Imitating progressive German and French institutions, the Ivy League schools had begun adding "natural philosophy" to their curriculums in the early 1800s, although they'd gotten off to a slow start. Yale's first natural history professor, Benjamin Silliman, Sr., had been forced to spend five months in Philadelphia, "then the scientific powerhouse of the nation," before he knew anything to teach. (The Philadelphians' habit of drinking wine and beer with meals had shocked him.) By the next generation, however, Yale's scientific faculty was well established. "Yale is determined to be up to the times . . . ," declared James D. Dana, professor of natural history, in 1856. "Why not have here, THE AMERICAN UNIVERSITY—where nature's laws will be taught in all their fullness, and intellectual culture reach its highest limit!?" George Peabody evidently didn't object to his nephew's ambition to learn nature's laws in all their fullness. He supported Othniel throughout the rest of his education, even when Aunt Judith Russell, George Peabody's sister, who was responsible for disbursing the nephew's allowance, complained about his sketchy accounts.

Marsh succeeded at Yale, and though eccentric, was well liked. His mineral collection grew so large that his landlord, who lived below his rooms, had to prop up the floor. The landlord's daughter recalled Marsh carrying her upstairs on his shoulders and offering to show her his minerals if she promised not to touch them. "He was always very odd and for most people it was like 'running against a pitchfork' to get acquainted with him," she said. His classmates called him "Daddy" or "Captain" because of his age, but admired his attainments. "One cannot look at him without thinking of Charles Kingsley's ideal naturalist," Charles H. Owen, a classmate, remarked, reeling off a Victorian list of outdoor virtues, including "strong in body, able to haul a dredge, climb a rock." Marsh graduated eighth in his class in the spring of 1860, and, among other honors, won a Berkeley scholarship for one to three years of graduate study at Yale.

The graduate promptly informed his uncle of his wish to study for "a professorship of natural science at Yale or some other college." Uncle George replied, "I heartily accede to what you propose," with the condition that Othniel "at all times treat your Aunt Russell with the utmost confidence with regard to your expenditures, as she will, as usual, pay them as authorized by me." Marsh chose the Sheffield School's Chemical Course, two years of classes in the various branches of chemistry, along with physics, geology, botany, modern languages, and the history of science. He published his first paper in 1861, on the Nova Scotia gold fields he had explored the previous summer. Although in marked contrast to Cope's precocity, the age of thirty does not seem too old for a graduate student's first publication. It evidently was too old for Othniel, however. He began to take a series of deliberate, careful steps that, within five years, would land him a position of power and security far beyond those of his collegiate contemporaries.

Marsh's work at Sheffield was distinguished enough for Yale to offer him a teaching post on his graduation, in 1862, but he declined it. He also sent his father a rare letter to complain of a report in the Lockport newspaper that Uncle George was sponsoring him for a Yale professorship. Such allegations might "do not a little in preventing me from getting it," he sagely commented. Instead, Marsh went to Europe, where he intended to study analytical chemistry and mineralogy in the fall, a move that took him closer to his London-based uncle and put him in position to act as go-between in some delicate negotiations being carried out by George Peabody and Yale. Uncle George, the greatest philanthropist of the mid-Victorian era, eventually gave away $8 million of the $12 million he amassed. One gift, of $2 million for low-income London housing, helped to keep England from recognizing the Confederacy in the Civil War.

Peabody planned a gift of $150,000 to Harvard. In late 1862, Othniel wrote to Professor Benjamin Silliman, Jr., in New Haven, informing him of this and adding that he had "strong hopes" that Yale might "yet be equally favored." Silliman enthusiastically responded that a like sum to Yale would pay for a science building whose donor's name was "sure to be always prominently and gratefully remembered." In May 1863, Marsh met his uncle in Hamburg to discuss a Yale donation and received a favorable response. At about the same time, he requested advice from Professor Silliman as to whether he should specialize in mineralogy, "which I should certainly prefer," or paleontology. Silliman advised him to choose paleontology, the only subject for which there might be a vacancy on the faculty. He also suggested that Marsh acquire an extensive fossil collection in Europe to bring back to Yale, adding that it would be desirable if Peabody could endow a chair in paleontology for his nephew.

In July 1863, Othniel wrote to Uncle George that a happy side effect of his "munificent donation to Yale" was a rise in his own hopes. "The faculty proposes to create a new professorship of geology and paleontology (the science of fossil remains) and give me the position" he wrote, and asked Peabody for three to four thousand dollars so that he could make the collection Silliman had suggested. "A library and cabinet is to a Professor of Science what capital is to a man of business," Marsh explained. He clearly knew whereof he spoke, since he was well on his way to gathering a lifetime supply not only of academic but of financial capital, without investing a dime of his own. It was a coup worthy of the subtlest financier. "By this appointment," he soberly gloated, "I shall at once be placed on a level with men all of greater age and experience than myself."

He already had acquired paleontological distinction with his *Eosaurus,* on which he'd published a paper in 1862. It was read to the London Geological Society by none other than Charles Lyell, the father of modern geology. He may also have met at that time the naturalist and Darwin-supporter Thomas Henry Huxley, an acquaintanceship that would prove invaluable. In the fall of 1863, he began attending paleontology lectures at German universities, delivered by the best authorities in the field. He had no trouble forming an influential circle of colleagues, "doubtless in part," Schuchert and LeVene observed, "an indication of the fashion in which his path was to be smoothed by the magic of George Peabody's name." His status as a beginner in a subject wherein he'd been offered a professorship may have seemed odd to his less upwardly mobile German counterparts, but Marsh quickly picked up fossil study, although he complained that he found zoology more interesting.

Evolutionary implications didn't trouble him. A dutiful churchgoer in youth, he never expressed religious sentiments in adulthood, and showed little interest in theory during his two years of paleontological training, when he concentrated on the logistics of finding and identifying fossils. At Solenhofen, where Cope marveled ambivalently at the bird-reptile *Archaeopteryx,* Marsh "added greatly to my collections . . . by good luck (and a week's hard work) I found a new fossil (articulate) which has interested me very much." His notes on lectures by Professor Ferdinand Roemer in Breslau were practical. "The most inviting field for Paleontology in North America is in the unsettled regions of the West," read one axiom. "It is not worthwhile to spend time in the thickly settled regions."

After a stay at Uncle George's Scottish castle and a visit to Charles Darwin's country home, perhaps arranged by Huxley, Othniel arrived back in New Haven in July 1865, with two and a half tons of fossils and books and, probably, with clear ideas about how he would spend the ensuing decades. In a way, those ideas would have been like the ones Edward Cope was entertaining in Haverford at the same time—of a life of research, collecting, and scientific correspondence. Yet as they unfolded, Marsh's plans would demonstrate an adamant linearity quite unlike Cope's diverse pursuits. They would be a New England stone wall beside the Bartram tradition's flowery Pennsylvania hedge. If Marsh was to collect professionally, for example, it would be fossils: period. "So much for mineralogy, which with me is now a reminiscence of the past," he wrote in 1863.

Othniel also seemed able to do without the emotional gratification and support that Cope found in marriage. This was partly good policy; Uncle George was strict about his protégés' marital affairs and might withdraw sponsorship at the hint of an undesirable match. Yet other Peabody protégés did marry. "It seems strange," wrote Schuchert and LeVene, "that such a fine matrimonial prospect should have been allowed to escape, and there are many indications of willingness on the part of various ladies to share his life. If romance really touched him, however, we have no hint of it." The only woman Marsh was known to have courted in his youth was Clemmie Dixon, daughter of U.S. Senator James Dixon of Hartford, and he evidently paid more attention to her mother, Elizabeth, who was important in Connecticut society, a good professional connection. "It must have been a sore disappointment to him that he did not become her son-in-law," Schuchert and LeVene wrote. "Just why he did not, does not appear from the letters we have. Aunt Eliza [Mrs. Dixon] had warned him, however, that the Dixon girls' feeling for him was respect rather than affection."

Marsh's careful planning didn't pay off immediately. Although Yale's faculty had unanimously approved his appointment ("a very unusual thing," he told Uncle George), the university had no money with which to endow it. Uncle George, piqued, suggested that part of his building donation be applied to endow his nephew's chair. Othniel wisely demurred, aware that Yale wanted every penny for the building, which would, after all, house his office and collections. It eventually became the Peabody Museum. Jobless, he stayed on the campus throughout the next year, sometimes giving unpaid lectures for indisposed professor friends. When the governing board in July 1866 finally voted to establish the paleontology chair, it was "without salary from existing funds," but Marsh took it without complaint. He was still getting an allowance from Uncle George, from whom he expected a legacy. An unpaid professorship also entailed no teaching duties or other distractions. Now the first professor of paleontology in the United States, Marsh could devote himself to becoming the country's leading paleontologist.

3

Fair Prospects in Dirt

Revolutionary War figures liked to say that they had fought for nationhood and that their sons (like George Peabody) had worked for the wealth to sustain it so that their grandsons could pursue higher things. Thomas Jefferson's remark along those lines certainly applied to Cope and Marsh: "We have spent the prime of our lives in procuring the precious blessing of Liberty. Let them spend theirs in showing that it is the great parent of science and virtue." Few people in history have begun their careers with the advantages of these two young men, and although they approached their post–Civil War positions with very different traditions and kinds of education, their lives must have seemed equally enviable. Not only were they financially independent; they had an almost untouched professional field spread out before them, the vertebrate paleontology of North America. It was a field not without difficulties and hazards, but the expansion of the country was making it more accessible every year.

What exactly was this profession? At a time when thousands were getting rich in the continent's gold and silver fields, why did a few become passionate about petrified bones? Paleontology has acquired a glamorous aura in the past century, but there's a certain superficiality about this. (The Jazz Age bonehunter Roy Chapman

Andrews was a model for Hollywood's Indiana Jones, but accountants made screenwriters change him to an archaeologist.) Compared with the cosmic explorations of physicists, a search for old bones can seem quaint and mundane, and it entails a lot of dirty work. Indeed, paleontologists are proud of their intimacy with dirt and the old-fashioned physical toil of moving it. A modern paleontologist, Spencer Lucas, curator of the New Mexico Museum of Natural History and Science, told me one reason he likes his science is that it "has technologically gone nowhere, and can't be done in cyberspace." Paleontologists seem to go into an almost trancelike state of communion with the dirt of fossil sites. At one site, I saw Brent Breithaupt, director of the University of Wyoming Geological Museum, pick up bits of fossil material and put them in his mouth. Asked whether he was tasting for something—the difference between petrified wood and bone, for example—he said no, he just did it to clean them. They all tasted like dirt.

My own amateur experience of paleontology has not been cosmic. Dirt prevailed when I helped to dig for some roughly 3.7-million-year-old, zebra-like horses at Idaho's Hagerman Fossil Beds National Monument. Edward Cope had discovered the species, *Equus simplicidens,* in Texas in 1892, but Hagerman has the world's greatest known accumulation of its skeletons. Although early diggers thought the horses died around water holes during a drought, the monument paleontologist Greg McDonald suspected that they drowned while crossing a flooded river during migration, as zebras drown in Africa. The excavation done in 1997 was to solve that problem, but my brief participation consisted of processing dirt. First I screened dirt back-hoed from the horse bed, feeling by day's end as though the dirt had been screened through me, although the work was enlivened occasionally by the fossils of small animals we were seeking. The most interesting was a muskrat jaw, an incongruous find, given the quarry's present location, high on a butte. Later, I scraped away dirt from the horse bed itself, which was less interesting, because I already expected horse bones. Even professionals seemed numbed after a day of this, and it was only the beginning of a three-month excavation. A small but noisy rattlesnake under the toolshed and gnat bites that itched and bled for weeks were the closest things to an Indiana Jones movie.

I found dirt prevalent even in paleontology's indoor, museum side. Preparing a roughly 55-million-year-old fish from Wyoming's Fossil Butte area (of a perch-like genus, *Priscacara,* also discovered by Cope) consisted of scratching petrified lake mud from the bones with a steel needle and then rubbing it off with an eraser. I scratched and rubbed for hours before I saw the first tiny bone, and exposing the whole roughly four-inch fish took weeks of intermittent toil. I was impressed that Cope had been able to name six *Priscacara* species from such bits of shiny brown fishbone. In the end, I felt as well acquainted with the powdery, petroleum-smelling lakebed dirt as with the fish.

Despite its *longueurs,* dirt-scratching is necessary for an understanding of paleontology. Doing it gave me a much stronger sense of what fossils are. And it's only part of the science. Paleontology does boast a modest technological sophistication. Greg McDonald planned to devise a three-dimensional map of the Hagerman horse bed with the latest computer technology, and to search for hidden fossils by using remote-sensing to pick up traces of uranium in the bones. (He later told me that at least two hundred horses had died in a stream channel. Instead of drowning in a flood, they'd probably starved in an almost-dry riverbed during a drought.) More important, paleontology has a respectably cosmic intellectual component. It is a much newer science than physics, and in the nineteenth century it was a more revolutionary one, producing changes in our understanding of time quite as profound as Galileo's were or Einstein's would be.

People didn't know quite what fossils are until around 1800. Although the ancient Greeks thought they might be the remains of long-dead organisms, most cultures regarded easily recognizable fossils, like giant vertebrate bones, as remains of mythical beings, and less obvious ones, such as shells or teeth, as minerals. Such objects, they thought, grew in the earth as crystals do, and any resemblance to living things was magical or occult. (The word *fossil,* derived from the Latin for "dug up," originally referred to any object dug from the ground, not just long-dead organisms.) After the Renaissance, naturalists began to look more closely at the resemblances. Leonardo da Vinci observed in the sixteenth century that shells in limestone were very like those of living mollusks in Italian mudflats. In the seventeenth century, the anatomist Steno saw a resemblance between the teeth of a large shark he was dissecting and fossils called "tongue stones." Still, such early naturalists didn't regard fossils as evidence that ancient organisms were different from living ones, or that life has taken myriad forms through millions of years.

An assumption that fossils reflect past faunas essentially like those of the present persisted through the eighteenth century. The naturalist Georges Louis LeClerc, Comte de Buffon, thought mastodon bones like those Charles Willson Peale displayed in his museum were remains of elephants that had lived in North America when the climate was warmer. The teeth were different from living elephants', but Buffon guessed that rhino teeth had become mixed with the elephant bones. Even stranger fossils were assumed to come from living species. When Jefferson obtained bones of an enormous unknown creature in 1797, he surmised they were from a lion that still roamed the unexplored American interior, and he named it *Megalonyx,* giant claw. Like most Enlightenment naturalists, Buffon and Jefferson considered extinction impossible, because it would have upset the balance of nature and other aspects of the divine plan. If extinction was impossible, then life always must have been essentially the same. Organisms like *Megalonyx,* which had vanished from one region, probably survived in another, Jefferson wrote, "for if one link in nature's chain might be lost, another and another might be lost, 'til the whole system of things should vanish by piecemeal."

Two French naturalists, Georges Cuvier and Jean Baptiste Lamarck, began to change this view of time in the early nineteenth century, when they studied piles of old bones and shells from Montmartre gypsum quarries. Respectively the foremost vertebrate and invertebrate anatomists of the day, they saw that the fossils were very different from living organisms, and that the past world must surely have been unlike the present. Cuvier in particular envisioned paleontology as a way of reconstructing vanished worlds through the study of anatomy. "I found myself as if placed in a charnel house," he wrote in an article translated in the *Edinburgh Review,* "surrounded by mutilated fragments of many hundred skeletons of more than twenty kinds of animals, piled confusedly around me. The task assigned me was to restore them all to their original positions. At the voice of comparative anatomy every bone and fragment of a bone resumed its place." As they studied fossils, the French naturalists realized that the apparently static present was just the surface of an unimaginably deep gulf of vanished worlds, and that living time does not simply pass; it changes, reeling through a constantly shifting series of landscapes and faunas. This temporal kaleidoscope is a commonplace now, but it might have seemed as bizarre to Thomas Jefferson in 1797 as the general theory of relativity.

There was "a grandeur in this view of life," to paraphrase Darwin, but it had a disagreeable corollary. The Enlightenment's belief that organisms could not become extinct no longer held, and the death not only of species, but of genera, families, and orders, joined that of individuals on life's tragic side. When Cuvier saw lithographs, in 1796, of a South American skeleton that resembled Jefferson's *Megalonyx,* he recognized its similarity to skeletons of much smaller, living tree sloths, so he classified the creature as a ground sloth, a mammal that apparently had vanished before civilization. And when Jefferson later saw Cuvier's article on this *Megatherium,* he realized that his *Megalonyx* wasn't a big cat but another ground sloth species. Lewis and Clark, whom he sent to the Pacific in 1804, found neither giant lions nor ground sloths living on the Great Plains.

Although Cuvier's and Lamarck's discoveries revealed an immense new dimension of time, they were not sufficient to explain how or why life changes. That unsolved problem set the two naturalists at odds. Lamarck thought extinct organisms had turned into new ones as their environment changed, and that life had an unbroken ancestry, leading back to its origins. He supported this theory of "transmutation" with the idea that organisms passed acquired characteristics to their offspring. If a land animal that lived beside a river took to swimming, for example, its offspring would inherit the ability to swim. Cuvier viewed such ideas as unscientific; he saw no evidence that old species had turned into new ones through heredity. When he examined mummified Egyptian animals several thousand years old, he found them identical with living ones, and the fossil record showed no continuum of organisms transmuting into different ones. "Lower" fossil organisms like mollusks did not merge gradually into "higher" ones like vertebrates. Organisms and groups of organisms appeared in the fossil record fully formed, and they disappeared without leaving apparent links to subsequent ones. To Cuvier, it was an unknown factor that had caused life's temporal kalcidoscope, not hereditary transmutation. And his view was considered the more scientific one until the publication, in 1859, of *The Origin of Species.*

Darwin's theory of transmutation through natural selection modified the Cuvier-Lamarck controversy but by no means ended it. Darwin showed how artificial selection of desired characteristics in dog- or pigeon-breeding could effect changes in domestic species. He suggested that a similar selective process operated on wild organisms, with those best fitted to their environment surviving to produce more offspring, thereby passing along their fitness. But he had trouble finding evidence of such progressive change in the fossil record. Cuvier's interpretation was about as valid in 1865 as it had been in 1829. Only a few vague and ambiguous "links" between fossil groups, like the birdlike creature that Cope saw in Solenhofen, pointed toward transmutation, and these were hardly enough to support Darwin's sweeping theory. Though Darwin maintained that the fossil record was too sparse to show the links, many naturalists thought that argument weak. The only way to resolve the controversy, it seemed, was to find and identify a great many more fossils.

Edward Cope and Othniel Marsh thus began their paleontological careers not only with financial independence and a near-virgin field of exploration, but at the cutting edge of contemporary science. If they could dig up enough fossils to demonstrate evolutionary transmutation and, perhaps, how transmutation occurred, they had a fair prospect of solving some of the century's major scientific problems. Brave new world.

Of course, it probably didn't seem that brave to them as they muddled through the postwar years. Marsh was determined to find fossils, but the idea that he would find many, much less important ones, must have been tenuous in 1866. Cope didn't, apparently, have much sense of a paleontological mission on his return from Europe. Herpetology always had been his strong suit, and perhaps he anticipated a distinguished career in that field. Indeed, he would have one. A commemorative biography published by the First World Congress of Herpetology, in 1989, called him America's greatest herpetologist, and added: "It has been said that Cope found herpetology an art and left it a science . . . Largely through his efforts it became a comprehensive entity including not only taxonomy but anatomy, geographic distribution, adaptation, and phylogenetic history."

Still, neither man wasted time getting into bonehunting, and their rivalry had already been established. In the fall of 1863, they had met in Berlin, where they may well have discussed paleontology, although the only description of the meeting is one that Marsh wrote in 1890, when he was trying to convince *New York Herald* readers that Cope was mentally and morally unfit. "Professor Cope called on me and with great frankness confided to me some of the many troubles that even then beset him," he wrote. "My sympathy was aroused, and although I had some doubts of his sanity, I gave him good advice and was willing to be his friend."

Self-serving as it is, Marsh's description gives an idea of how the meeting went. Cope doubtless did most of the talking, bragging of his accomplishments while complaining about his problems. He must, however, have been impressed by Marsh's degrees, scholarships, and calm, methodical careerism. Encountering the highly qualified Marsh would have been a shock, given the genteel amateurism of his own education. And wily Marsh certainly would have taken Cope's measure as an associate or a competitor. He may have sensed advantage in the bug-eyed Quaker's loquacity and felt a twinge of envy at hearing of Cope's wide knowledge and freedom from academic entanglements. It's hard to guess what Marsh's "good advice" may have been, but it's easy to believe that both felt a nascent rivalry beneath their amicable manners. Their disparate backgrounds predisposed them to look down, subtly, on each other. The patrician Edward may have considered Marsh not quite a gentleman. The academic Othniel probably regarded Cope as not quite a professional.

In his *Herald* article, Marsh added that he'd seen Cope often during the next five years "and retained friendly relations with him although at times his eccentricity of conduct, to use no stronger term, was hard to bear." Their letters to each other mention the meetings in passing, and one gets the impression of a judicious older man dealing cautiously with an impetuous younger one. Cope's letters are scribbled on all sides of the sheet, sometimes on top of earlier lines, and lurch from boasts of new discoveries to complaints about Philadelphia Academy elders to puppyish avowals of friendship. Marsh's are legible and cordial, if a bit pompous. The two men remained at least outwardly friendly through the decade, as shown by their scientific compliments of naming fossil species after each other. Yet fundamental discords surfaced inexorably. It was one of their 1860s meetings, according to Cope, that started the trouble. According to Marsh, it was another.

Cope's scientific activities after his return from Europe are strenuous just to read about. He explored Catskill caves in 1864, collected Ohio Paleozoic amphibians in 1865, found a Virginia fossil whale in 1866, and went to West Virginia in 1867 for more cave exploration and some metal prospecting. His most exciting find was at

Haddonfield, New Jersey, a marshy area underlain by soft, limy marl deposited in the Cretaceous Period, the late dinosaur age. Farmers had been uncovering huge bones there since the 1830s, and in 1858 Joseph Leidy, Cope's Philadelphia anatomy instructor, had named one of the first American dinosaurs, *Hadrosaurus,* from a Haddonfield skeleton. In August 1866, Edward wrote to his father that he'd received a letter from the superintendent of the West Jersey Marl Company about some giant bones. "I thought best to go down at once as the letter was old, and accordingly hired a carriage at Camden next morning. I found the remains of much greater interest than I had anticipated—being nothing more nor less than a totally new gigantic carnivorous dinosaurian probably of Buckland's genus *Megalosaurus*! Which was the devourer and destroyer of Leidy's *Hadrosaurus,* and of all else it could lay claws on . . . It is altogether the finest discovery I have yet made—apart from those resulting from indoor study."

The English anatomist Richard Owen had coined the name *dinosaurs* in 1841, but the "terrible lizards" remained little known. Owen thought the giant saurians had walked on all fours, like reptilian rhinos. Impressed by the shortness of their front legs, however, Leidy and Cope surmised that they'd walked on their hind legs or, perhaps, had hopped like huge kangaroos, and in 1868 they helped the British artist Waterhouse Hawkins build the first skeletal reconstruction of a bipedal dinosaur at the Academy of Sciences.

"This carnivore, then, is an interesting link between those of the mammalian series, and the carnivorous birds," Cope wrote in a newly established journal, *The American Naturalist.* "If he were warm blooded, as Prof. Owen supposes the Dinosauria to have been, he undoubtedly had more expression than his modern reptilian prototypes possess. He no doubt had the usual activity and vivacity which distinguishes the warm blooded from the cold blooded vertebrates." The new saurian was a striking find for a twenty-six-year-old paleontologist. Cope soon dropped his original attribution to the English genus *Megalosaurus* and named the fossil *Laelaps,* after the goddess Diana's hunting dog, which was magically turned to stone while in mid-leap. Naming an extinct giant must have made him feel godlike, and evidently such experiences so enthralled him that he quit his teaching job in 1867. He'd been having a disagreement with the faculty and administration, anyway. "Flummery there is and will be at Haverford," he wrote to his father, " 'til it is put out of the question by legislation."

Leaving pedagogical flummery behind, Edward set out to devote himself to science. "It was a happy, busy time," wrote Henry Fairfield Osborn. "His wife and little daughter were in splendid health . . . His only concern was the need for money, not for family expenses, as he was able to afford two maids, but that 'I may have something to spend on scientific work.' " There was plenty of work. In the year after he left his professorship, while still living in Haverford, Cope not only wrote several papers on Haddonfield dinosaurs but published "his first complete synopsis of the extinct Amphibia of the world."

According to Cope, Marsh entered this Eden as a serpent. In March 1868, Edward wrote to his father: "After the work of the winter, I have been taking a little outdoor exercise as it has come conveniently in the shape of a trip through the Marl country. I had intended making it this season, but not so early. My friend, Prof. Marsh of Yale College, had however planned to go a little earlier, so I accompanied him . . . Prof. Marsh has studied and traveled in Europe for three years and is very familiar with their invertebrate fossils." The pair found "three new species of Saurians . . . tho' the weather has been very bad" and Cope's "liver had been disturbed." It sounded like a friendly jaunt, though there was a hint of Marsh the outsider, the invertebrate expert, descending on a busy, tired Cope and dragging him through the mire. In 1890, however, Cope drew a woeful picture of the visit's consequences. "I took him through New Jersey and showed him the localities," he complained in the *New York Herald*. "Soon after, in endeavoring to obtain fossils from those localities, I found everything closed to me and pledged to Marsh for money considerations."

Cope made Marsh sound like *Laelaps aquilunguis,* "devourer and destroyer . . . of all . . . it could lay claws on," in the *Herald.* "During my early acquaintance with Professor Marsh," he continued, "I offered him every facility to examine my own collection and gave him all information regarding localities of fossils that I knew of . . . At that time Marsh never published anything without consulting me —without, in fact, getting my judgment on what should be published and how it should be published. He invariably followed my advice and gave me no credit. I found at length that he was using me . . . to furnish him with brains. I left him and since then have suffered a steady persecution for over twenty years."

Marsh's 1890 *Herald* version of their initial quarrel differed completely. He wrote cryptically that he'd tolerated the younger man's eccentricities until "the number was approaching the Biblical limit of seventy times seven, when a break occurred between us, and since then we have not been friends." The break's cause was not the Haddonfield visit; it was a colossal piece of anatomical stupidity on Cope's part. Marsh told the *Herald* that in 1869 Cope had restored the skeleton of a giant, long-necked marine reptile, a plesiosaur named *Elasmosaurus,* with the skull attached to the end of the tail. "The skeleton itself was arranged at the Museum of the Philadelphia Academy of Sciences, according to this restoration," wrote Marsh, "and when Professor Cope showed it to me and explained its peculiarities, I noticed that the articulations of the vertebrae were reversed and suggested to him gently that he had the whole thing wrong end foremost. His indignation was great, and he asserted in strong language that he had studied the animal for many months and ought at least to know one end from another." Joseph Leidy then corrected the mistake, Marsh concluded, and "when I informed Professor Cope of it, his wounded vanity received a shock from which it has never recovered, and he has since been my bitter enemy."

Memory and self-interest evidently distorted both 1890 *Herald* versions of the feud's beginning. Cope's mistake with the *Elasmosaurus* skull probably was not as stupid as Marsh made it sound.

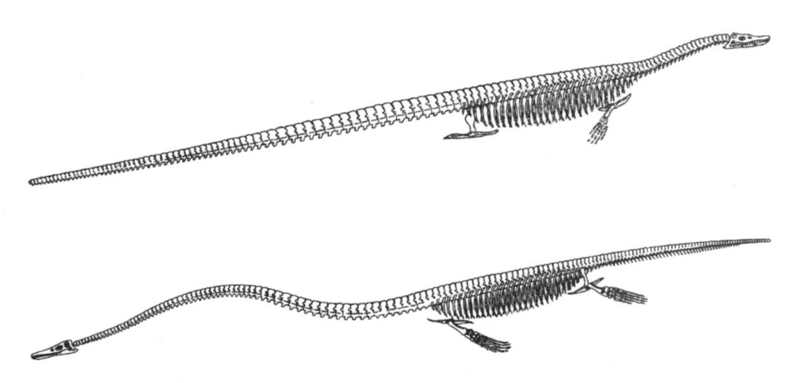

Cope's Elasmosaurus, *before and after*

Cope explained in the *Herald* that he had based his backward reconstruction on an earlier, equally erroneous one by "another scientist" (the unnamed Joseph Leidy), and that "my predecessor saw the error he and I had committed and corrected it in print without previously informing me." On the other hand, Marsh also accused Cope of paleontological claim-jumping during the period—that is, of writing papers on fossils that had been "promised" to him—so Edward's attitude of injured innocence before Othniel the Usurper is suspect. Neither man seems to have "broken" with the other during that time. Their letters to each other in the late 1860s mention both affairs without particular animosity. Marsh named a fossil after Cope, *Mosasaurus copeanus,* in 1869, which would have been a strange thing to do if he'd just usurped Cope's Haddonfield fossils. After moving his household to Haddonfield in 1867 or 1868, Cope lived there for the next seven years, also strange behavior if Marsh had frozen him out of the local fossil market.

Far from separating, Cope and Marsh spent the late 1860s becoming more entangled. Cope evidently was as much attracted by Marsh as repelled, because he grew more and more like him in that period. He jettisoned any pretense of Jeffersonian amateurism, and, despite leaving his university post, aggressively pursued his counterpart's professionalism. "I have occupation enough to last me for years," he told his father in 1869, "but my opportunities of doing it are much diminished by my capital being locked up in another business . . . Thee objects to my selling the farm, and I have had respect to thy opinion as being founded on experience and caution, yet can not see why a man who is not a farmer and is never likely to be should get his living from that source." Edward soon sold McShag's Pinnacle, completing his metamorphosis from land-based naturalist to capitalized scientist.

Already near the top of the professional ladder, Marsh, older and calmer, had less outward reason to be influenced by Cope, though Edward's accusations of fossil-and-idea-theft suggest that Othniel secretly lacked confidence. Schuchert and LeVene wrote that he "did not at once turn definitely into the channel that he was to follow for the rest of his life" after getting his Yale professorship. He frequented museums, where he talked to curators, but the only other place beside Haddonfield where he did much on-site collecting was the Connecticut Valley; there, he acquired a number of dinosaur tracks between 1867 and 1871. His published papers, mainly on Paleozoic invertebrates, were modest compared with Cope's. Less scatterbrained than the younger man, Marsh also was less energetic, and Cope must have seemed an unnerving jack-in-the-box, springing up wherever fossils were available—Virginia, Ohio, the Carolinas. Marsh already was inclined to buy his fossils instead of digging them up himself, while Cope would remain an enthusiastic digger throughout his life.

Spencer Lucas described an affair that seems typical of Cope-Marsh interactions at the time. A Philadelphia missionary, Dr. Elias Root Beadle, was an enthusiastic collector, and in 1867 acquired the skull of a *Lystrosaurus,* a mammal-like reptile from the South African Triassic Period. Aware of Marsh's position, he wrote to Yale asking whether Marsh wanted to see it, but he apparently got no answer. Cope, however, was interested in Beadle's fossil, and in 1870 he described and named it in the *Proceedings of the American Philosophical Society.* "We cannot help wonder," Lucas wrote, "if Marsh, who was a keen student of fossil reptiles, felt some chagrin at having missed the opportunity given to Cope to describe the skull of *Lystrosaurus.* Perhaps a positive response to Beadle's 1867 letter would have brought the skull to New Haven for Marsh to study." In fact, Marsh finally did buy the skull in May 1871. "Why Marsh was so interested in purchasing a specimen that had already been described is unclear," Lucas observed, "but we may imagine that he carefully checked the accuracy of Cope's description and illustrations of the specimen."

Cope's energy may have pushed Marsh to move faster and farther than he would otherwise have done, although the Professor doubtless would have denied any Philadelphia influence. New England was on the move in the nineteenth century. His father's brother John had gone to California in the 1830s, and young Othniel and Caleb almost joined him in the gold fields in 1850. Whatever the reason, Marsh took a giant step a few months after the March 1868 Haddonfield trip with his rival. He finally applied the "axiom" he'd copied in his Breslau lecture notes four years earlier, and in August left the "thickly inhabited regions" in which it was "not worthwhile to spend time," and moved into the "unsettled regions" of North America, "the most inviting field for paleontology."

4

Professor Marsh's Traveling Bone and Pony Show

Colonists had been finding fossils in North America since a man named Samuel Maverick reported shells and bones under Virginia's James River in 1636. Cotton Mather regarded a mastodon tooth, found near Albany in 1705, as evidence of antediluvian giants, and pious Yankees a century later interpreted the Connecticut Valley's dinosaur tracks as the footprints of the raven sent forth by Noah. Yet fossils were rarities in the damp forests east of the Mississippi. Vegetation hid them, and roots and rain soon destroyed those exposed by erosion. Only under special circumstances were large deposits accessible. Jefferson's *Megalonyx* lay under a cave floor, and mastodon bones like those in the Peale Museum largely came from bogs, marl beds, and similar morasses, where tannin, salts, or other chemicals had preserved them. The few fossils found in the eastern part of the continent didn't provide compelling support for Cuvier's and Lamarck's tremendous vision of time. With some exceptions, like those from Haddonfield, the bones were relatively recent ones found in surface deposits, and the Pleistocene epoch species in them usually resembled living animals.

When naturalists began venturing west of the Mississippi, however, they met strikingly different conditions. Weather and vegetation on the high, dry prairies toward the 100th meridian are much kinder to fossil-hunters than in eastern forests. Petrified bones last longer, because grassland soils are less acid than forest ones, and grass roots lack the crushing power of tree roots. They are easier to find because wind and water erosion erase the grass in places, forming blowouts and gullies that have eaten into the soft plains rocks for centuries, eventually creating "gullies" miles across and hundreds of feet deep: the badlands. This erosion has exposed fossils that have been accumulating in various sedimentary strata across the West for hundreds of millions of years.

Aspiring paleontologists couldn't have asked for a better representation of the Cuvier-Lamarck vision of the past than the American West. Ferdinand Roemer, the Breslau professor from whom Marsh had acquired his "go West" axiom, had seen this in the 1840s when he was unearthing Paleozoic trilobites on the Texas prairie. Some badlands lay along main trade routes, like the White River region in the Dakota Nation between Fort Pierre and Fort Laramie, now the site of Badlands National Park. "From the uniform, monotonous prairie," the geologist John Evans wrote about that area, "the traveler suddenly descends one or two hundred feet into a valley that looks as though it had sunk away from the surrounding world . . . Embedded in the debris lie strewn, in the greatest profusion, organic relicts of extinct animals." As traffic expanded, travelers brought east more and more of these "organic relicts."

In 1846, a fur trader sent part of a jaw with some big molar teeth from the White River Badlands to Hiram Prout, a physician in St. Louis. Familiar with Cuvier's work, Prout attributed the teeth to a Montmartre gypsum genus called *Palaeotherium,* an early relative of tapirs and rhinos. Some of his other badlands fossils came to Joseph Leidy's attention in 1851. Leidy, born in Philadelphia in 1823, was mid-nineteenth-century America's best comparative anatomist, a relic of his city's fading eminence. A recent biography called him "the premier representative of the old school of eighteenth- and early nineteenth-century naturalists." He saw the similarity between the badlands fossil and Cuvier's fossil, and he named this creature

Palaeotherium proutti. But he also realized that it was a much bigger animal than the French one. It eventually was renamed *Titanotherium.*

Although he had been trained as a doctor, Leidy was such a modest and retiring man that his medical practice was short-lived. It had taken him six weeks to recover from his first human dissection, and he never adjusted to dealing with live patients. He thus had plenty of time to examine fossil bones; indeed, they came to obsess him. "You can have no idea how much my mind has become inflamed on this subject," he wrote to his fellow naturalist Spencer Baird in 1851. "Night after night I dream of strange forms: Eocene crania with recent eyes in them." During the following decades, he went on to name dozens of strange Western skeletons.

"He soon learned that he could not base his study of American fossils on the work of French paleontologists," wrote Henry Fairfield Osborn, "for the life of our Western regions was not known in the Old World. Every specimen represented a new species or a new genus of a new family, or in some cases a new order." Osborn admired Leidy as a "quiet, unpretentious, unassuming wonderfully gifted observer of nature . . . in the exact spirit of Cuvier," and called his book, *The Extinct Mammalian Fauna of Dakota and Nebraska,* "the finest contribution that has been made to vertebrate paleontology in this country, if not the world." Leidy is regarded as America's first professional paleontologist, although he himself harked back to the "amateur" Bartram tradition when he wrote, in the preface, that his book was simply "a record of facts . . . as the author has been able to view them. No attempt has been made at generalizations or theories which might attract the momentary attention and admiration of the scientific community."

Leidy's scholarly modesty caused him a major frustration in one area, however. Since teaching was his only source of income, he was unable to go West to study the badlands fossils *in situ* during the 1850s and 1860s. He depended on explorers like John Evans to acquire them. "For whom are you now collecting?" he wrote plaintively to another young explorer-naturalist, Ferdinand V. Hayden, in 1854. "If I am to be of service to you, or can in any way promote your objects, let me know it." Hayden proved a good contact, and in the following year he sent Leidy fragments of Cretaceous Period dinosaur bone, from the Montana badlands, that resembled the Haddonfield fauna.

As it happened, Leidy had good reason to shy away from the West. Until the 1850s, naturalists had been able to travel across the Great Plains in small groups or alone without too much danger. The botanist Thomas Nuttall had used his rifle mainly to dig up plant specimens when he wandered as far west as the Platte River in 1811. Acquisitive Indian bands might rob unwary travelers, but the tribes felt secure in their possession of the country and were tolerant of well-behaved white visitors. Lone observers like George Catlin could watch their picturesque lives with an idealism little changed from William Bartram's. This idyll faded, however, as white intrusions grew, and the plains became much less safe after 1854, when a Lakota band wiped out twenty-nine soldiers led by a foolishly belligerent officer near Fort Laramie. The army retaliated the next summer by killing eighty-six Lakota men, women, and children at Blue Water Creek in Nebraska, a slaughter that set off the intermittent thirty-five-year-war ending at Wounded Knee, south of the White River Badlands.

Just when Indian hostility punctured romanticism about the West, general Euro-American attitudes toward remote places were beginning to change. Romantic observers had emphasized "sublimity" in their descriptions of the wilderness—strangeness and picturesqueness. Victorian observers instead stressed hardships and dangers, in part because newspapers like the *New York Herald* found that sensational accounts of narrow escapes or grisly disasters sold better than the old "sublime" style. At every chance, the *Herald* ran stories on Indian violence, often with gruesome and lubricious details—grotesque scalpings, immigrant girls raped and then dragged naked through the sagebrush behind their captors' ponies. The Bennetts probably had no particular feelings against Indians. They seem not to have traveled much west of New Jersey.

Yet westward expansionism was more popular than Native American culture, to say the least, and bloodthirsty savages made better copy for their urban readership than did the ambiguous conflicts of the real plains.

Leidy's supplier, Ferdinand Hayden, demonstrated the shift from sublime to sensational while describing the White River Badlands in an appendix to Lieutenant Gouverneur Warren's report on the Blue Water Creek fight, a report itself so grisly that the army tried to suppress it. Hayden began by echoing the romantic convention of likening the buttes to fabulous cities: "It reminded me of what I imagined of the amphitheater of Rome, only nature works on a far grander scale than man . . . All around us were naked, whitened walls, with now and then a conical pyramid standing alone." Then he lapsed into a more lurid, ghoulish vein: "We felt very much as though we were in a sepulcher, and indeed we were in a cemetary [sic] of a pre-Adamite age, for all around us at the base of those walls and pyramids were heads and tails, and fragments of the same, of species not known to exist at the present day. We spent that day and the following exploring the cemetary, which the denuding power of water had laid open for our inspection."

Hayden's macabre vision was prophetic of the bison bones and human corpses that soon would litter the plains, and that vision seemed to linger when I visited the South Unit of Badlands National Park in 1997. The May morning was bright, but barely a tuft of grass grew on the mudstone bluffs, and the sun bleached out the earth colors admired by modernists like Frank Lloyd Wright. I didn't see any "heads and tails"—a century of collecting had done for them—but the tortured slopes were full of large holes that looked so much like lairs that I fancied the "heads and tails" might have come to life and crept away underground. The watchful holes reminded me of a diorama I'd seen the day before—at Agate Fossil Beds National Monument in Nebraska—of a Miocene epoch carnivore named *Daphoenedon* emerging from a burrow to challenge a huge, vaguely piglike *Dinohyas* for possession of rhino carcasses scattered in a drought-shrunken pond. As though the subject wasn't gruesome enough, the squabbling prehistoric animals had been re-created as skeletons, making the diorama an evolutionary *danse macabre.* The only animals shown whole were giant vultures, circling down from a hectic sky.

Of course, the Badlands National Park holes weren't ghostly burrows, but their real nature was macabre enough, as I'd learned a few days earlier from Spencer Lucas in New Mexico. He had shown me how such holes open into deep pits carved in the soft mudstone by rain and wind. A veneer of dried mud can conceal a pit just as ice hides glacial crevasses, and Lucas told me he'd once fallen, up to his waist, in a twenty-foot-deep pit so narrow that, if he'd dropped to the bottom, he'd have become a fossil paleontologist.

I poked around the South Unit's watchful holes with unease as well as fascination, and my unease grew when a car full of Lakotas stopped and pointedly asked whether I was "fossil-hunting." The South Unit is on the Pine Ridge Sioux Reservation, and fossil theft has made Indians proprietary of their skeletal fauna. I said I wasn't, truthfully enough, but when they departed, I realized I'd left some Kansas chalk on the floor of my car down the road. Imagining myself hauled into the reservation police station to answer for five centuries of land grabs, I glanced at the Indians' tire tracks and saw the only fossil bone I would encounter that day, a leg joint. It vaguely resembled a deer's, but was a beautiful creamy yellowish stone, and probably came from an oreodont, a sheep-size ungulate of the Oligocene epoch, twenty million years before deer reached North America. I'll never know, though, because I left it there.

In 1857, a "very large force of the Dakotas" met a Black Hills expedition on which Ferdinand Hayden was serving and refused to let it proceed, lest it disturb the bison they depended on for their winter food and robe supply. "Their feelings toward us, under the circumstances," wrote the expedition commander, Gouverneur Warren, "were not unlike what we should feel toward a person who should insist on setting fire to our barns." Warren sensibly turned back, so Hayden didn't get to see much of the Black Hills that year. He continued to explore the upper Missouri River region safely through the late 1850s, perhaps because the Indians thought his obsession with giant bones a sign of harmless lunacy. Western exploration's romantic phase ended, however, when "Picks-up-stones-running" (supposedly Hayden's Lakota name) joined the Union Army at the Civil War's outbreak.

Young Professor Marsh faced a daunting situation when he set out for the postwar "unsettled regions." Not even bonehunters could hope for special tolerance by 1868, a year after Colonel George Armstrong Custer's cavalry had massacred a sleeping Cheyenne Village on Washita Creek. Othniel C. Marsh was no romantic, however, and he didn't let the grimness deter him; his first trip west was a model of cautious practicality. He joined an excursion party from the American Association of Science's Chicago annual meeting in early August. The scientists rode the new Union Pacific Railroad to its terminus at Benton, Wyoming. Then they rode back—and Marsh managed to augment his scientific capital even from this tentative foray.

There had been newspaper stories of "fossil human remains" along with tiger and elephant bones dug from a well at a place called Antelope Station in Nebraska. A correspondent with the byline "H.J.W." had written in the *New York Times* on June 22 that the bones were "undoubtedly human" and that they "would attract the attention of geologists and scientific men. For if it be correct that human bones are found . . . in the tertiary, I suppose there will have to be some revision of current scientific opinion both as to the origin of the human race and the date of its existence upon this planet." Such ancient human remains might have buttressed pre-Darwinian ideas of life's history, since they suggested that humans had been living unevolved since the "creation" of the Tertiary Period, the age of mammals.

Marsh's professional expertise served him well, however, and he made a quick meal of the newspaperman's "undoubtedly human" bones. "Before we approached the small station," he wrote in an unfinished memoir, *Fossil Hunting in the Rocky Mountains,* "I made friends with the conductor and persuaded him to hold the train long enough for me to glance over the earth thrown out of this well . . . I soon found many fragments and a number of entire bones, not of man but of horses, diminutive indeed, but true equine ancestors . . . Other fragments told of his contemporaries—a camel, a pig, and a turtle at least."

The conductor curtailed Marsh's stop, but he had paid the stationmaster to set aside the fossils. "A hatful of bones was my reward when I passed Antelope Station returning east," he wrote, "for the station master had kept his promise. As we shook hands, I left in his palm glittering coin of the realm . . . A hasty examination of my new fossil treasures, as I showed them to my fellow travelers, disclosed no trace of man, but positive evidence of no less than eleven inferior animals, all extinct . . . Recalling the old adage that 'truth lies at the bottom of a well,' I could only wonder, if such scientific truths as I had now obtained were concealed in a single well, what untold treasure must be in the whole Rocky Mountain region. This thought promised rich rewards to the enthusiastic explorer in this new field, and my own life work seemed laid out before me." Employing similar lapidary imagery, his biographer Charles Schuchert wrote that the little three-toed horse, *Equus parvulus,* was "the first in a long series of fossil horse remains that was to form one of the chief jewels in Marsh's crown." Marsh quickly converted his treasures into scientific currency; he presented a paper on them to the National Academy of Sciences convention in late August, and published it in the November issue of the *American Journal of Science.*

As if this wasn't enough to wipe Edward Cope's eye, Othniel distinguished himself in herpetology during his lightning prairie tour. He returned with specimens of mysterious gilled salamanders that lived in Lake Como near Medicine Bow, Wyoming. The Smithsonian herpetologist Spencer Baird had given these "devil fish" the species name *Siredon lichenoides,* but biologists suspected that they were overgrown larvae of an already known terrestrial species, the tiger salamander, maybe prevented from maturing by the stressful conditions of their alkaline lake. When Marsh got them back to Yale, most of the specimens obligingly lost their gills and climbed out of the water—tiger salamanders. Publishing the find along with his "human remains" piece in the *American Journal of Science,* he wrote that they "apparently belonged to the species *Amblystoma mavortium* Baird, as recently defined by Prof. Cope in his able review of the Amblystomidae," and thanked "his friend E. D. Cope for various suggestions." Considering the deteriorating state of their relations, it's hard to say whether this was a sop or a taunt. Maybe it was a bit of both. In an 1868 letter, Cope maintained that Marsh had "designed for him" a pair of the *Siredons,* but there's no evidence that he got them. Marsh did send photos, for which Cope thanked him brusquely, ending: "excuse haste this time."

Marsh immediately started planning a return to the West. Uncle George Peabody died in 1869, leaving him $150,000, so there would be no more accounting to Aunt Russell. "Imminence of Indian Wars" prevented a trip that year, however. A new approach to Western exploration clearly was required, one that would surround the fossil collector with a reassuringly large retinue. Fortunately, Othniel was developing a flair for publicity. His devil fish discovery and the "human remains" debunking had received wide newspaper play, even though not all of it was positive. On August 29, a *New York Times* piece on the National Academy of Sciences convention credited him with proving "by personal inspection" the nonhuman nature of the Antelope Station bones, but on the same day, the *New York Herald* had slighted his convention paper in an editorial fulminating against professional science: "A good many peddlers in science have been holding conventions of late. There has been a convention at Northampton, Mass., where . . . they discussed such objects as . . . human bones found at Antelope Station on the Pacific Railroad, and other subjects equally abtruse . . . The subjects discussed at these scientific conventions may be very entertaining to the muddleheads collected there, but they are of very little service to the public."

Then, in late 1869, Marsh chanced on an affair that changed his image to that of a public servant as well as a shrewd investigator. Well-diggers on a farm at Cardiff, a hamlet in upstate New York's

Onondaga Valley, uncovered a ten-foot "petrified" humanoid that religious fundamentalists hailed as fossil evidence of biblical giants. A delegation of geologists who examined the figure said it was no fossil, but they were impressed anyway. "Altogether it is the most remarkable object brought to light in this country," said James Hall, the New York State geologist. "Although not dating back to the stone age, it is nevertheless deserving of the attention of geologists." The farm's owner, a Mr. Newell, made a small fortune showing the "giant" before he sold it to a cartel of professional exhibitors.

Newspapers also hailed the find. On November 18, the *Herald* ran a long article on "the Onondaga Giant," turgidly calling it "a wonder that will perhaps long remain a mystery as well, for the antiquarian ready to entertain and welcome anything that promises a clue to the ancient world, as for the man of exact science, too eager sometimes . . . to reduce all evidences to the levels of the positive evidences of his day . . . It seems for the present evident that the so-called giant is an unmistakable antiquity." The article mainly dwelt on the giant's significance as evidence of a vanished New World civilization, but it included remarks by a Professor J. J. Brown, who believed it was a cast of a real human body, like the relics at Pompeii. The *Herald* writer mused that the "intensely realistic character of the figure" argued against its being a sculpture.

This was only a year after James Gordon Bennett, Jr., had taken over the *Herald,* and he doubtless was involved in the story. He'd been quick to put his mark on the paper. Born in 1841, he had spent most of his childhood in France, because his mother had fled there to escape the social ostracism incurred by his scabrous father. ("Petticoats, petticoats, petticoats!" a Bennett Sr. editorial had taunted New York society. "There you fastidious fools, vent your mawkishness on that.") After his return to America, society had embraced the teenage multimillionaire as passionately as it had rejected his father. "Jimmy Bennett would serve as exhibit A in any treatise on the fluidity of the American caste system," wrote the historian Richard Kluger; "in a single generation the central figure of

James Gordon Bennett, Jr.

the family was lifted from pariah to grandee." Jimmy had blossomed into a playboy whose antics—among them, nude buggy rides—included a sport called "Bennetting," which involved galloping, Cossack-style, into student outings from a young ladies' boarding school near the New York Union Club. His response to expulsion from Newport's Reading Room for riding a pony into its exclusive precincts was to establish another club, across the street from his own villa. For exercise, he rode a bicycle around his Fifth Avenue block while a servant waited on the curb with a silver tray holding brandy.

After his appointment as editor in 1867, Jimmy's first act was to pull Bennett Sr.'s name from the masthead, at which his enraged parent rushed downtown to reinstate himself as proprietor. James Sr. soon realized he had a worthy heir, however, and made Jimmy publisher the next year. Ability, energy, and an iron will lay under the playboy's curly blond hair. Richard O'Connor wrote that the *Herald* "was undoubtedly the most brilliantly edited, enterprising, and talked-about newspaper in the world" in the years following Jimmy's "ascendancy to the management." Oswald G. Villard, later editor of *The Nation,* recalled that rival journalists didn't dare to go to bed before seeing the *Herald*'s early edition, "which they picked up in fear and trembling lest they find in it one of those record-breaking stories which made its name as famous as that of 'The Thunderer' [the *Times* of London]." Bennett Jr. sometimes *didn't* go to bed; he worked on his paper around the clock and generally slept four hours a night.

Jimmy's journalistic abilities fell below his father's in only one respect. Bennett Sr. had been a paragon of terse, hard-hitting prose, but his champagne-brained scion was given to bombast and verbosity. The tortuous description of the Onondaga Giant in the November 18 *Herald* may have been a sample of his prose, or at least his editing. The August 1868 editorial attacking the National Academy of Sciences convention was well written, and its philistine disdain for useless scientific "muddleheads" was more like Bennett *père* than *fils*. Young Jimmy didn't care whether something was useless as long as it amused him, and digging up ancient objects evidently did, to a degree. Only a few weeks before the antiscientist editorial, the *Herald* ran one supporting a fossil museum in Central Park and advocating that "special instructions . . . be issued to our Indian agents and the commanders of our frontier garrisons to encourage not only the collection of fossil remains but of living animals."

Whether or not he had worked on the Onondaga Giant article, the *Herald*'s proud new proprietor would have been nettled on November 29, when the *Buffalo Courier* and *Syracuse Journal* printed a letter from a "man of exact science" who had examined the giant. That scientist, Professor O. C. Marsh, stated peremptorily that it was a fake; there were conspicuous tool marks on its surface, which ground water would have dissolved from the highly soluble gypsum if it had been buried for more than a few years. The letter was "at once widely copied by other newspapers throughout the country," wrote Schuchert, "thus exploding what has been called . . . the most uproarious hoax ever launched upon the credulity of a humbug-loving people."

Actually, Marsh was not the only giant debunker. On November 21, the *Herald* itself had carried a brief editorial to the effect that a "Professor Boynton," who originally had accepted the giant's antiquity, was "beginning to think, from experiments he has made on the action of water upon gypsum," that the bones had not been in the ground longer than a year. Musing that the hoaxers would have done better had they put the giant in "some old not much visited Indian mound, and had planted some good sized trees," the editorial concluded that Boynton "doubtless" would be proved correct. Marsh's remarks, though, had an abrasive confidence that would have grated even on Bennett's arrogance. "I was allowed to make a more careful examination of the statue than is permitted to most visitors, and a very few minutes sufficed to satisfy me that it is of very recent origin, and a most decided humbug," the newly installed Yale Professor had written. "Altogether, the work is well calculated to impose upon the general public, but I am surprised that any scientific observers should not have at once detected the unmistakable evidence against its antiquity."

The affair must have been a vexing lesson to young Jimmy on the pitfalls of journalistic credulity. The *Herald* "copied" Marsh's *Buffalo Courier* letter on December 1, but its attitude was stiff: "We are permitted to publish a letter by him to a friend, containing his views thereon," the piece began tartly. "From such a source, opinions are entitled to be of great weight on such a subject and it must be admitted that the testimony of Professor Marsh finally settles the claim of the monstrosity to be of ancient origin." Another giant-related *Herald* item the same day seemed to corroborate Marsh, yet subtly undermined him. Marsh had assumed that the giant's gypsum was from New York State, but Bennett reprinted an article from an Iowa paper stating that the figure was made of Iowa gypsum, "as shown by the analysis of Doctor———." The *Herald* buried both reprints in its back pages, and on December 10 it muddied the waters with affidavits from the *Albany Journal* in which the giant's discoverers swore to its authenticity. Bennett evidently had no further space for the Yale Professor's "weight" that year, although he extensively covered Uncle George Peabody's death and large funeral.

It turned out that Newell's brother-in-law, a free-thinker named George Hull, had made and buried the figure in 1868 to embarrass biblical fundamentalists, so Marsh's cocksure debunking prevailed, despite the *Herald*'s sour coverage. He became a new kind of media figure, the saturnine Professor who penetrates humbug with steely gaze and esoteric knowledge. "Fortunately, he was not only one of the most eminent of living paleontologists," wrote Andrew White, president of Cornell University, who had been chagrined by his colleagues' gullibility, "but, unlike most who had given an opinion, he really knew something about sculpture, for he was familiar with the best galleries of the old world." Fifty years earlier, Georges Cuvier had wowed French journalists with his flair for resurrecting extinct animals. "Our immortal naturalist has reconstructed worlds from blanched bones," wrote that prolific hack, Honoré de Balzac. Marsh expanded on Cuvier, popularizing the paleontologist's deductive skills to synthesize a scientific precursor to Sherlock Holmes.

What red-blooded young man wouldn't have wanted to go West with such an expert? As the summer of 1870 approached, many a Yale student with a thirst for adventure and a family willing to finance the trip wanted to go. Marsh had one requirement for the new exploration: a self-supporting workforce. He picked thirteen stalwarts (including the grandson of the cotton gin inventor Eli Whitney) and gave them a shopping list: "trusty pistols and the all-essential knife, as well as the geological hammer and the appliances of science." The second requirement—actual protection from Indians for the Professor and his blue-blooded coolies—was less ready to hand, but Marsh's reputation fortunately had impressed the U.S. Army, too.

"That he found favor in their eyes is not strange," wrote Charles Schuchert; "he was preeminently an outdoor man in those early years, a crack shot, a fisherman of repute, a seasoned camper, at his best around a campfire where the swapping of tall tales is a highly appreciated art." Generals Philip Sheridan and E. O. C. Ord became Marsh cronies, and General William Tecumseh Sherman provided him with a letter "which became an open sesame at all army posts." The generals knew good publicity when they saw it. Two student participants wrote accounts of the first Yale expedition for Eastern periodicals: Charles Betts in *Harper's New Monthly Magazine* and Harry Ziegler in the *New York Herald.* Both gave the army high marks. Ziegler likened the expedition to Julius Caesar's campaigns: "They, too, went, saw, and conquered."

There's no evidence as to how the *Herald* came to employ Ziegler, but it's not surprising that it did. Its owner's love-hate relationship with explorers was in its first flush, and young Jimmy would have looked forward with relish to the haughty Yale Professor's adventures in the land of the Sioux and Cheyennes. Unwary travelers were being scalped in full view of the new transcontinental railroad's stations at the time, so there was no telling what might happen to Othniel and his boys.

The party of student cavaliers, Fifth Cavalry, and Pawnee guides that clanked and jingled out of Fort McPherson, Nebraska Territory, in early July of 1870 might have been an early version of Buffalo Bill Cody's Wild West Show. In fact, Buffalo Bill himself went along as one of two white scouts, although only for a day. That was enough for Bill to add the Professor to his own repertoire. He expressed skepticism about Marsh's learned avowals that the Rocky Mountains had once been a shallow sea and the Nebraska plains a lake, adding that in the evening the Professor "tipped me a wink as much as to say, 'you know how it is yourself, Bill!'" Cody later visited Marsh when his show played New Haven, and Schuchert wrote that "it was probably a fine advertisement for show and museum alike when he rode up to the ornate iron gates of the latter with long locks flowing out from under his ten-gallon hat." (It was probably less fine when Bill touched the Professor for $80 during an 1874 tour, though he did pay it back.)

Cody displayed the moccasins of an Indian recently killed near the fort, so the clanking and jingling may have seemed justified at first. Charles Betts wrote that the Pawnee scouts, "with movements characteristic of their wary race, crept up each high bluff, and from behind a bunch of grass peered over the top for signs of hostile savages." But the Sioux and Cheyennes failed to perform. Prairie fires were the only "signs," and horses bitten by rattlesnakes or shot accidentally were the only casualties. In lieu of frontier gallantries, the expedition performed a lot of fossil-collecting, perhaps more than its members had anticipated. Betts reported that the temperature often stood at 110 degrees in the shade, an amenity available only under wagons, since the prairie lacked trees or even rocks. "After fourteen hours in the saddle," he wrote, "one of the soldiers, exhausted with heat and thirst, finally exclaimed, 'What *did* God Almighty make such a country as this for?' 'Why,' replied another more devout trooper, 'God Almighty made the country good enough, but it's this deuced geology the professor talks about that spoiled it all!' "

Marsh himself might have made a good general if poor eyesight hadn't kept him out of the Civil War. In the next six months, the Yale party went on many bonehunting forays, looping over wide swathes of rugged terrain without getting lost or running out of supplies, although the students sometimes overate the commissary. The July trip from Fort McPherson crossed the northwest Nebraska sandhills to Loup Fork Creek, where fossils included those of six horse species and two rhinos. In August, the party moved to Colorado and found an outcropping of the same fossils that characterize the White River Badlands, including Joseph Leidy's titanosaurs as well as rhinos, giant tortoises, and oreodonts. They then went north to Antelope Station, Marsh's 1868 "human remains" site, where they added three more horse species to *Equus parvulus*. In September, escorted by the 13th Infantry and a Mexican guide, they went west to Fort Bridger in southwest Wyoming.

Hayden had sent Leidy a few fossils from north of Fort Bridger, but the area to the south was almost unknown. "No exploration of this region had ever been made," wrote Betts, "but Indians and hunters had brought back stories of valleys strewn with giant petrified bones." The party crossed the Uintah plateau and followed the

Green River to its confluence with the White, where "a grand scene burst upon us. Fifteen hundred feet below us lay the beds of another great Tertiary lake. We stood upon the brink of a vast basin so desolate, wild, and broken, so lifeless and silent, that it seemed like the ruins of the world." The Bridger area would prove very important during the next few years, but the 1870 Marsh party found mainly fish and small mammal fossils, not giant bones. They also ran into horse thieves, who appropriated a grain cache. Marsh went into their hideout to retrieve it. "Endeavoring to control his embarrassment by speaking as to ordinary ranchmen," wrote Betts, "our illustrious chief remarked blandly, 'Well, where are your squaws?' 'Sir,' replied a dignified ruffian, 'this crowd is virtuous.' "

Marsh gave his charges a rest in October. They spent two weeks sightseeing around Salt Lake City, flirted with "twenty-two daughters of Brigham Young in a box at the theater," and then, "overcome by the effort, immediately crossed the Sierra Nevada to San Francisco." After visiting the Mother Lode and Yosemite, they returned by rail to a Green River locality in Wyoming and went back to work, collecting fish and insect fossils. The season's final foray was from Fort Wallace in western Kansas, where Marsh wanted to work Cretaceous marine beds along the Smoky Hill River. There he found a climactic fossil that would allow him to expand his professorial persona to the fullest.

Huge bones of the marine lizards called mosasaurs were common, but a smaller one proved more important. Marsh spotted it one evening while riding along an old buffalo trail beside the river. He wrote in his memoirs that it was hollow, "about six inches long and one inch in diameter, with one end perfect and containing a peculiar joint I had never seen before." Back in New Haven, the only similar bone he could find was that of the flying reptile called the pterodactyl, but all the pterodactyl bones then known were tiny. "After further comparison, however, I could only return to the dragon's wing for the nearest resemblance," he wrote, "and believing in my science as taught by Cuvier, I determined to make a scientific announcement of what the fossil indicated, and trust to future discoveries to prove whether I was right or wrong. I therefore made a careful calculation of how large a Pterodactyl must be to have a wing finger corresponding to the fragment I had found, and ascertained that its spread of wings would be about twenty feet." Marsh hoped to prove his daring deduction by finding more of the giant pterodactyl in the next fossil-hunting season.

The Kansas prairie also provided climactic True West adventures. The Smoky Hill area was an Indian war flashpoint because of the railroad, and when their horses and mules stampeded one night, the Yale party members feared an attack. It was just coyotes trying to steal some fresh meat, but the panicked lifestock kept going until they returned to Fort Wallace, where the post commander also became alarmed and sent a rescue detachment racing to Marsh's camp. There, they found the expedition placidly cooking Thanksgiving dinner, which they invited the soldiers to share. "Yale songs were sung by our party," Marsh wrote, "and western stories were told by the army officers. The November wind howled through our camp, and coyotes, sniffing the feast, serenaded us from the orchestra bluff above, but we heeded them not, for we were safe and happy."

The southern bison herd still existed in 1870, even though the animal was doomed by the transcontinental railroad. Marsh told a bison story that, according to Schuchert, established "his reputation, once and for all, as a hunter," but that sounds as much about power as sport. Like Buffalo Bill, frontier types tended to be skeptical of the pistol-packing professor, so Othniel felt bound to prove himself. "On arriving at the table land where the buffalo were to be found," he wrote, "great numbers were in sight, mainly in two vast herds . . . Captain R. suggested that the best place for us to see the hunt was from the ambulance . . . With this advice, he mounted his horse and joined the rest of the escort, giving general directions to the ambulance driver and other teamsters to keep to the rear."

"Captain R." was a well-known frontier skeptic, and Marsh clearly hated being sidelined by such a figure. Eager to shoot three nearby bulls, he "rebelled" and bribed the ambulance driver to take him after them. He said excitedly, "Driver, do you want a Five Dollar bill?" and the driver replied, "Mighty Bad." After a dash across the prairie, Marsh killed one bull and liked the feeling so much that he paid the driver ten dollars more to help him kill the other two, despite the terrified mules' frenzied rearing. As he descended from the ambulance to approach one of his prizes, the driver suddenly exclaimed, "Look out, he's going for you!" and the bull got up and charged. Marsh dodged and shot it down, however, and started to butcher it for the head and hide. Captain R. rode up at that point, to Marsh's satisfaction.

"He had seen the chase after buffalo with the ambulance, and, looking at the mules, reproved the driver," he wrote. "I took all the blame, and covered with blood as I was, with my hunting knife in hand, he was polite enough not to reprove me, and even helped me load my gory trappings in the ambulance." The driver, Marsh continued, "tipped me a sly wink at the time, and touching his vest pocket where three greenbacks were safely deposited, he sat silent except to urge on his tired mules."

Marsh seemed to know just how far such adventures should go. The expedition had suffered neither serious illness nor major injury when it returned to Yale on December 18, an extraordinary display of logistics. "Its success was far-reaching," Schuchert wrote. "Directly, it gave to the museum thirty-six boxes of material that included many an important specimen destined to become the type of a new genus or species. Indirectly, the wide publicity it received from newspapers in the East and in the West focused attention both on the market for fossil vertebrates and on Professor Marsh as the most active figure in that market."

A week later, Harry Ziegler's *Herald* piece presented "the most active figure" with a high media gloss. "The Professor of Paleontology . . . in Yale College is O. C. Marsh, a young, fresh-looking gentleman of probably five and thirty years of age, who appears to the casual observer anything else than a devoted student of the petrified bones of past ages," it began. "But if one could obtain a glance at the professor's cabinets, in College Street, he would everywhere see not only the evidences of great scientific acquirements, but valuable specimens of great scientific treasures." Ziegler piled on the gilt, adding that Marsh was curator of the Yale Geological Museum, and that he would "busy himself arranging the treasures he brought back" as soon as he could "perform the thousand offices of duty and friendship which awaited him on his return from his summer's explorations."

Ziegler even dredged up Othniel's 1869 debunking of the "human remains" at Antelope Station. "If Professor Marsh saw me write this I am sure he would suggest at this point some reflections of the injury done to science by the hasty opinion of the New York doctor who pronounced fossil horses and hoofs human remains," he declared in tortuous *Herald* style, "and I am not sure but that he would suggest also an admonition for the newspaper correspondent—one of the famous writers on the war—who told the world so monstrous a fiction upon an authority so slight."

Jimmy Bennett would have relished this slur on a "hasty opinion" in the rival *New York Times,* although the expedition's dearth of scalpings may have disappointed him. The article was not slavish in its adoration of the professorial idol. A skeptical tone crept into its account of Kansas "sea serpent" fossils. "In spite of Buffalo Bill's disbelief in his 'shallow sea' theory, Professor Marsh was intent on finding evidence of what the famous frontiersman would have considered a bigger snake story than the others," the *Herald* remarked, "but since his return all that he has been heard to say about his snake discoveries is that 'the investigations of the party were crowned with success.'" This passage departs surprisingly from the scholarly effusions of earlier ones, and the tone of subsequent sentences was even more mocking: "The vertebra of the sea snakes in the Yale Museum show the kind of success with which their investigations were crowned, and we may expect in the scientific journals articles of learned length and thundering sound similar to those which Professor Marsh has already published . . . In science as well as in fiction a well-connected tale can be built upon very slight foundations. Professor Marsh has a single joint of a venerable sea serpent's backbone from which he will be able to construct a snake not less than sixty feet in length."

The article ended with a come-on, promising more professorial adventures. "It is understood that Professor Marsh has not yet satisfied himself that he has obtained everything of value to science or new in the fields of scientific research which some of these regions afford, and that he will return to the Rocky Mountains to explore further the great fresh water lake basins of that interesting locality." In fact, Ziegler went West with Marsh again the next summer, which suggests that a sequel was planned. Bennett's paper, however, would not further extoll Othniel's Western exploits. Maybe the 1871 advent of African and polar explorations eclipsed the bloodless Yale expeditions. Maybe dissipated Jimmy had heard enough about the "young, fresh-looking gentleman." (In August of that year, he ran a lengthy editorial attacking the Yale faculty as having "no learning.") The Professor would shine again in the *Herald* before long, but in a different, more complex role.

5

The Lone Philadelphian

Perhaps stung by the rude skepticism that had crept into his *Herald* profile, Marsh was eager to confirm his giant pterodactyl deduction. "I confess that during the long winter . . . I at times had my misgivings, at least as to the size the animal should have had," he wrote, "and I longed for spring to come." In June 1871, the second Yale expedition hurried straight to Fort Wallace and the Smoky Hill River. Harry Ziegler was the only repeater, but Marsh had manned the trip easily, and it was well publicized again. The *New York Times* covered it in a fall 1871 article, and a graduate student named Oscar Harger described it for the Yale *College Courant*. Harger was a genius who had put himself through Yale by juggling statistics for insurance companies, but he dropped his mathematical career for zoology. Marsh paid his expenses, $950, and hired him as assistant for the next year. He would work for Marsh until his early death, in 1887, and would play a posthumous part in the *Herald* scandal.

They found the prairie as miserably hot as it had been cold in November, with temperatures up to 120 degrees, but the Professor got his pterodactyl. "As soon as the tents were pitched," he wrote in his memoirs, "I started with two or three companions to seek the locality." Marsh quickly came upon a cross he had cut in the earth in November, and a moment later "was at the spot where I discovered the fossil bone, and soon detected fragments of it lying near, partially covered up by the loose chalk that had been washed over it by the winter storms." He found the rest of the original bone and then most of the other wingbones. "These I measured roughly as I took them up one by one," he exulted, "and . . . I was soon able to determine that my calculations based on the fragments were essentially correct, and that my first found American dragon was as large as my fancy had painted him. My journey from New Haven was amply repaid, but greater rewards were to come, for during the month that I spent at hard work in this region, other dragons came to light, even more gigantic and much more wonderful than I had before imagined." Marsh showed that his pterodactyls had wingspans of up to twenty-five feet. So much for the *Herald*'s "well-connected tales built on very slight foundations."

The parched expedition fled Kansas in August, and moved to the Green River Basin near Fort Bridger, where they stayed for six weeks. They found abundant small skeletons, and shipped eleven boxes of them east, before passing to a new locality in eastern Oregon, the John Day Basin. This Tertiary deposit contained the remains of elephants, rhinoceroses, lions, and other "tropical animals," as well as several species of fossil horses. After collecting another eleven boxes, they visited Portland and San Francisco, and then some members returned east by rail. Marsh and the others sailed back via Panama, where he started a hobby of collecting pre-Columbian artifacts.

"From a scientific point of view, our trip has been a very successful one," concluded the *Times* correspondent (although he devoted more space to the sensational topics of Indians and Mormons than to fossils). Marsh may have had an unpleasant surprise when he reached Yale in mid-January, however. He no longer was the only professional paleontologist working the West. While he'd been combing the Green Basin, Jack-in-the-box Cope had popped up on the Smoky Hill River.

Cope's motives for going west are not as well documented as Marsh's, but they're not hard to surmise. He would have heard plenty about Yale's wonderful 1870 expedition. He also would have envied Joseph Leidy's exhaustive new book, *Extinct Mammalian Fauna of Dakota and Nebraska.* "The vast extent of our country west of the Mississippi seems to have been the arena on which were enacted . . . some of the most important events in the geological history of the American continent," said the introduction by Ferdinand Hayden. "There are indications that in this region are still to be wrought out some of the most important problems in geological science." Stuck in Haddonfield, Cope had to compete with his former teacher for the fossils that explorers were sending east. He'd felt guilty pleasure when a review had faulted Leidy for his reluctance to theorize and had expressed hope that a younger man, perhaps "the accomplished . . . discoverer of Laelaps," might someday "elucidate" Leidy's fossils. "Not handsome or Christian," he had written his father, "yet I cannot help feeling some gratification, as it does not equal in unhandsomeness the manner in which . . . Leidy . . . treated me." The review was signed only by the letter H, but Edward suspected that its author was Thomas Henry Huxley, becoming widely respected as an authority on paleontology.

There was only one way to top Leidy: go west, especially since Leidy didn't seem prepared to go there. But at this point Cope encountered his noninstitutional background's limitations. Marsh clearly had found an effective way to deal with the Indians and other logistical problems, but it was not a way open to a lone Philadelphian. The second Yale expedition had cost $15,000, then an astronomical sum. He would have to find another way.

Getting appointed to one of the geological surveys that had sprung up after the war was the most promising way of going there safely, and it had the added advantage of being published in survey reports. After failing to get appointed to a Texas survey, Edward turned his attention to his Philadelphia acquaintance Ferdinand Hayden. Despite the West's mounting dangers and a University of Pennsylvania geology professorship, Hayden had continued to explore and had worked his way up in the nascent federal scientific bureaucracy. By 1871 he was head of the Geological and Geographical Survey of the Territories, with a growing staff and an annual budget approaching $100,000. "Building on the reputation he had established before the war," wrote his biographer, Mike Foster, "Hayden did his job so well and made such effective use of publicity that more than anyone else he drew attention to the importance of western surveying."

Although a New Englander raised in New York, like Marsh, Hayden had gravitated to Philadelphians like Joseph Leidy because of a desire to distinguish himself as an explorer-naturalist on the Bartram model. His career was a last public efflorescence of traditional multidisciplinary natural history before specialization prevailed. "He wanted to discover the mysteries of the Great American West —all of them," Foster wrote. "These different aspects added up to a wholeness, an intangible totality the uniqueness of which could only be appreciated by studying all the aspects together." By the late 1860s, "Hayden was the star spokesman of the unified approach, and because of his commanding prestige and the renown of his survey, that approach ruled the thinking of his day."

Hayden also was a romantic traditionalist, in his dewy-eyed optimism about the frontier. "Surely the great West, with its broad fertile acres, to be had almost for the asking, through the generosity of our Government, is the poor man's paradise," he wrote, and he espoused theories that increased rainfall would "follow the plow" and permit a dense population. Such optimism was ill-suited to the West, but it was popular with Western legislators, immigrants, and restless Eastern farmers, and it attracted congressional appropriations. "The growing financial support of government for science," Foster wrote, "was a direct result of the huge enthusiasm Hayden stirred up for his broadbrush approach to natural history and of his popular approach to the West." Some of the Eastern press also got into the act. The exploration-minded *New York Herald* was one of the papers that ran admiring accounts of Hayden's summer expeditions.

Cope got Hayden to appoint him to his survey as a paleontologist in the spring of 1871, but making something tangible out of that was another matter. "I never can rely on him," Edward complained to his father. A straightforward explorer, Hayden was, of necessity, a devious bureaucrat as he navigated the treacherous shoals of government funding. Marsh, having served on a Smithsonian panel that renewed Hayden's appointment for the Nebraska survey, had influence of his own. No salary or other funds pursuant to Cope's appointment had materialized by late summer, and impatience evidently overcame him. In September Edward went west on his own, in an excited mood.

"I am now in a country that interests me greatly and the prospects are that I will be able to do something in my favorite line of vertebrate paleontology," he wrote to his wife from Topeka, Kansas, on September 6. "I had for the first time a view of the prairies, and they are wonderful to me and look more like ocean than anything I have seen." A week later, he was on the west Kansas high plains, where "the train ran close by a herd of buffalo! . . . There were only 30 or 40 and the bulls tried to cross ahead of the engine, but became alarmed and turned back. They cantered along close to the cars, a splendid sight . . . About the fort they appear in countless thousands; indeed, there are millions they say between here and Denver. Carcasses and bones lie on each side of the R.R. track, from Bunker Hill to this place; remains of animals shot from the car windows."

Since the Smoky Hill River fossils already were familiar in the East, Cope planned to spend only a week there before following Marsh's trail to the little-known beds around Fort Bridger. "Marsh has been doing a great deal, I find," he wrote, "but has left more for me, and one of his guides is at Fort Wallace, *left behind,* and in want of a job. Prof. Mudge wanted to accompany Marsh, and Marsh wouldn't let him go. I'll let him go!" His attempt to hire Mudge failed, however, perhaps because Marsh already had let his attitude toward his colleague be known. Edward also couldn't get a railroad pass beyond Fort Wallace, so his season's collecting ended there.

Luckily, Marsh's visits had educated the Fort Wallace military in fossil-hunting without prejudicing them against rival paleontologists. "I am very fortunate in getting here at this time," Cope wrote on September 21. "Now the men and officers of this regiment (5th Infantry) know the country well, and can guide us as to localities. They have been with Marsh and Mudge and know where they have been and can save my going over the same ground . . . Indeed the stories I hear of what Marsh and others have found is something wonderful, and I can now tell my own stories, which for the time I have been here are not bad."

Cope got off to a slow start. His first fossil hunt, though it covered "a circuit of twenty miles," located "few bones." I could understand this when I was in western Kansas; it's a big place. The fossil-bearing chalk bluffs are widely scattered, and I saw no recognizable bones in a blistering day of searching the Monument Rocks east of Fort Wallace. Glowing yellow against the dark-blue prairie sky, the chalk contained a lot of organic material but nothing I could decipher, and even the paleontologist Greg McDonald was uncertain about the samples I showed him. He was fairly sure grayish veins were fossil oyster beds, less sure that a reddish cylindrical clot was a coprolite, a fossil fece. He couldn't say of what, except that it seemed too small to be a mosasaur's.

I was there in relatively lush May. In late summer, Edward had to go miles to find silty, alkaline water, and sometimes he had to dig for bones while wearing a handkerchief tied over his face to shield his eyes from windblown dust. There were more serious concerns. As Marsh already had discovered, western Kansas was perhaps the main area of Indian conflict at the time, since the Cheyennes and Arapahos were infuriated by the wanton bison slaughter along the railroad. Nothing remains of Fort Wallace but scraps of metal in a local museum, but bullet holes from attacks are still visible on the railroad buildings from the 1860s. Twenty miles from the fort, a war party could have made short work of a few civilians and a squad of infantry. In July, according to the *Herald,* an Oklahoma Cheyenne band had burned two teamsters at the stake.

Cope soon got the hang of collecting on the prairie, however. "If the explorer searches the bottoms of rainwashes and ravines," he recalled, "he will doubtless come upon the fragments of a tooth or jaw, and generally find a line of such pieces leading to an elevated position on the bank or bluff where lies the skeleton of some monster of the ancient sea. He may find . . . a pair of jaws lined with horrid teeth, which grin despair on enemies they are helpless to resist." Edward reported that he'd found remains of mosasaurs, plesiosaurs, sea turtles, giant tarponlike fish, and—not to be outdone by Yale—pterodactyls.

"The flying saurians are pretty well known from the descriptions of foreign authors," he wrote. "Our Mesozoic periods had been thought to lack these singular forms until Professor Marsh and the writer discovered remains of species in Kansas chalk. Though these are not numerous, their size was formidable. One of them (*Pterodactylis occidentalis,* Marsh) spread eighteen feet, between the tips of the wings, while the *P. umbrosus,* Cope, covered nearly twenty-five feet with his expanse."

Cope's twenty-five-foot pterodactyl must have annoyed Marsh, but another of Edward's fossils would have amused him. The most "wonderful" story Cope told was of finding the skull of a "huge Mosasaurus" projecting from the bank of a ravine. He and his assistants spent nearly two days digging out the rest of the skeleton, with some help from their military escort. He decided that the spine and tail went through an entire spur of the ravine, because the tip of the tail came out on the other side. He excitedly estimated that the "animal must have been 75 feet long," even though they'd been unable to get it all out. A seventy-five-foot mosasaur would have been a "very rare animal" indeed—most of the giant marine lizards were no more than half that long. In fact, the tail on the other side of the bank belonged to another mosasaur skeleton.

If Cope lacked his rival's meticulousness in excavating and reconstructing bones, his paleontology was more imaginative. He had a sense of natural diversity, which, though a product of his eighteenth-century education, seems more modern now, more environmentalist, than Marsh's professorial sportsmanship. Everything interested Cope; he saw ecosystems. "The flowers we have often heard of," he wrote to his sister from Kansas, "but I did not suppose they were so tall; as high as a man's head or a steer's back. Sunflowers, cornflowers and various composites, flax, sage, euphorbia, endless verbenas, etc."

Sometimes Cope wrote with the self-dramatization Marsh displayed in his memoirs, but often it was in a simple spirit of wonder. "Prairie dogs and antelopes are in thousands and the wolves howled around us as we worked the bluffs," said a letter written to his wife, Annie, on September 21. "Coyotes and Jackass rabbits abound, skunks are more common than all, and very beautiful. I will try to get a prairie dog for Julia." To Julia, his five-year-old daughter, he wrote: "Thee mustn't be too sure about the prairie dog yet a while, as I am not certain whether I can get one or not. But I will try. This animal is not a kind of dog really, but is a little like a fat squirrel with a short tail. They live in great numbers together, each family having a hole in the ground."

Cope extended his eye for ecological diversity from the living prairie to the dead one. Like Leidy, he envisioned his prehistoric subjects in an almost hallucinatory way. One field assistant, a Kansan named Charles H. Sternberg, recalled that when his boss "began to speak of the wonderful animals of the earth, those of long ago and those of today, so absorbed did he become in his subject that he talked on as if to himself, looking straight ahead and rarely turning toward me, while I listened entranced." More than the methodical Leidy, who had yet to see a badland, his visions included environments as well as organisms.

Edward saw the Cretaceous midcontinent's shallow seas as a detailed living world and was one of the first to imagine the geological past as such. "These strange creatures flapped their leathery wings over the waves," he wrote of giant pterodactyls, "and, often plunging, seized many an unsuspecting fish; or, soaring at a safe distance, viewed the sports and combats of the more powerful saurians at sea. At night-fall, we may imagine them trooping to the shore, and suspending themselves to the cliffs by the claw-bearing fingers of their wing-limbs." He saw not only individual species, but associations of them. "Many other huge reptiles and fishes peopled both sea and land . . . Far out on the expanses of this ancient ocean might be seen a huge, snake-like form, which rose above the surface, and stood erect, with tapering throat and arrow-shaped head, or swayed about, describing a circle of twenty feet radius above the water. This was the *Elasmosaurus platyurus,* Cope, a carnivorous sea reptile." If any one American naturalist was the inspiration for the prehistoric images that have captured the twentieth century's imagination, it was Cope. His rudimentary drawings of them have an odd liveliness despite their crudeness.

Cope may have exaggerated the success of his first collecting trip. It was necessarily on a much smaller scale than Marsh's. If the Professor's populous adventures resembled Buffalo Bill's Wild West Show, Cope's were like William S. Hart films—tales of self-reliant, taciturn wanderers. Edward was not taciturn, of course, but he was self-reliant. He would arrive by rail or stage at a town near some badland, hire a wagon and crew, and ride out in search of fossils. At Fort Wallace he had the assistance of an army detachment, but usually his crews were frontier riffraff. His style was less efficient than Marsh's, and he got no publicity, a serious political handicap. Although highbrow papers sometimes carried his name in their general science reportage, they ignored his low-key expeditions. Not even the Hayden-loving *Herald* noticed them, since they operated apart from main survey parties. Yet Cope's forays were surprisingly productive. His visions of past life seemed to impell him toward its fossilized relics.

Cope's trips were like Marsh's, however, in that both men were sound managers, careful about expedition planning and logistics, even though both liked to take calculated risks and make daredevil noises. Quaker Edward ostentatiously carried no pistols, and when the Fort Wallace soldiers tried to lend him some, "left them behind as I hate the sight of them." But he always was well equipped with hunting weapons, since living off the land was economical. Neither man would have succeeded without this combination of bravado and caution, which contributed both to their reputations and their survival.

Edward clearly enjoyed his first trip despite its hardships. His exhilarated letters home from Kansas show that he was determined to go back to the West as soon as possible, no matter how eager he was to see his wife and daughter. After a brief visit to Denver, he returned to Haddonfield in late October with a very respectable bone haul. He hadn't brought back anywhere near thirty-six boxes of fossils; still, he must have felt reassured that imagination and determination could compensate for a scarcity of funds and institutional support in gleaning a share of Western treasures. His next expedition would be less reassuring.

6

Babel at Fort Bridger

The fossil beds around Fort Bridger were a paleontological gold mine, and Cope was eager to get there. "The history of the life of the successive ages of Earth's history has nowhere a greater prospect of elucidation," he wrote to Commander E. O. C. Ord, regional army head. Western Kansas's already well-known marine reptiles weren't a great scientific challenge just because they were so different from living organisms. A vanished world of sea monsters supported Cuvier's conventional view that life had changed through abrupt, wholesale discontinuities.

The Fort Bridger beds dated from the Eocene epoch, the early age of mammals, and although the fossils that Hayden and Marsh had found there were strange, they bore a similarity to living organisms that supported the Darwinian gradual transmutation thesis. Joseph Leidy had identified some bones as looking like those of still-living lemurs, which are primitive primates related to monkeys and humans. Fossils of fish, crocodilians, freshwater turtles, and plants resembled many organisms now living in North America. The area held unparalleled promise for the paleontological bread and butter of finding and naming fossils as well as for the more refined delights of theorizing, and Cope must have been elated when he managed to badger Hayden into sponsoring his trip there in the summer of 1872.

The Bridger beds were so tempting that Joseph Leidy finally went West that summer, at the invitation of a local physician, James V. Carter, who had found the area's first fossils in 1868. Carter also had befriended Marsh during his 1870–1871 expeditions, and suggested that he accompany Leidy to Fort Bridger in 1872. "I would be delighted to have you together here," he wrote to him in June, "and my faith is that you will work beneficially to all in this way. Dr. Leidy is a genuine friend of yours and I believe a most generous man . . . Now I say, can't you come alone—that is, not with an 'expedition.' I think a great deal of you, Professor, but with due respect I don't 'go a cent' on the Yale boys as helpers to science."

Marsh wasn't having any such unstructured arrangements, however, and he didn't welcome the prospect of Cope or Leidy being in what he considered his territory. According to his colleague C. E. Beecher, he "entered every field of acquisition with the dominating ambition to obtain everything that was in it, and leave not a single scrap behind." To him, other collectors were going after what was rightfully his. He feared not only the loss of fossils, but the loss of credit if others described and named fossil species first, which may have been likely, given his slow methodical approach. According to Henry Fairfield Osborn, he "disputed Cope's right to enter the Bridger field and put every obstacle in his way." Edward himself charged, in a letter to a scientific friend, that Marsh had tried to get Hayden to promise to keep his geological survey out of the Bridger area, promises "which were never made." Marsh, therefore—as he wrote in an 1873 *American Naturalist* article—asked Cope and Leidy to agree to an arrangement whereby they would send each other dated copies of any articles they published on Bridger fossils.

This arrangement sounds equitable, especially if all three paleontologists spent that summer in the field where primitive conditions hampered publishing. Marsh, however, faulted Cope for not cooperating with it. Yet Marsh wasn't in the field during the summer of 1872. According to his 1873 article, he spent the time "between July 22 and October 8" at Yale, publishing "a series of fourteen papers on vertebrate fossils of the West." He "started west" on October 8, "when the last paper of my series was published." This delay seems so incongruous with his dog-in-the-manger attitude toward the Bridger that many writers have assumed he did go West that summer. This was another manifestation of his shrewdness. Othniel would write his papers in comfortable New Haven while Leidy and Cope stumbled around where, for two seasons, he already had deployed a small army of collectors.

Cope certainly did stumble. The younger man sounded confident when he left Haddonfield in mid-June. "I will have every facility furnished by the Interior Dept," he wrote to his father, "expenses paid; orders for men; wagons, beasts, provisions etc." He also had promising assistants: a schoolteacher named Samuel Garman, with whom he'd been corresponding, and two boys he'd met in Chicago. Hayden still wouldn't give him a salary, but that at least freed him from government discipline. His confidence faded when he reached Fort Bridger late in the month.

Hayden had taken all the available mounts on an expedition to Yellowstone, so Cope had to hang around until mid-July before he could get an outfit (his letter to Commander Ord was a belated attempt to ingratiate himself with the army), and he had to pay for it himself. James Carter wrote to Marsh that Cope had been there some time but had "done nothing." His putative assistant proved self-serving and unreliable. "This man, Garman, had already received from me an equivalent of at least $125 and learned enough to enable him to collect well himself," Cope later told his father. "He passed himself off as *Friend* but I suspect this to be false, and his whole scheme was to get up an expedition of his own." The post personnel did not welcome him as had the men at Fort Wallace. Marsh had seen to that.

Marsh's own local bonehunters, whom he'd employed during his trips, were worse than unfriendly to Cope. While waiting to get his outfit together, Cope passed the time by making short trips to the fossil fields around the fort. One day he saw a Marsh team dig at a spot and then depart without taking any bones. When he went to the site, he found part of a skull with a few teeth; the shattered quality had evidently discouraged the Marsh party. Cocky Edward, though, was not discouraged; he took the fossils home and named a new species for them. Years later, his paleontologist friend W. B. Scott expressed doubt about Cope's species and showed him one of his own specimens. It had the same skull but different teeth. After comparing the two, Cope admitted, "You're right; those fellows fooled me."

Fort Bridger is now a state historic site, and some of its original frame buildings remain, looking very tiny against the surrounding peaks and badlands. Its isolation in 1872 was profound, and Cope's heart must have been in his mouth when he finally started into the wilderness on July 16. Because Samuel Garman had made "ridiculous and unreasonable demands of pay for services, time, etc. and poisoned the two boys with mutinous ideas," he had fired them, reducing his staff to a cook, a packer, a teamster, and a garrulous local who called himself "Sam Smith of the Rocky Mountains." Smith was the only man in Fort Bridger who was willing to guide the party, and Edward ingenuously described him as "first rate."

Probably hoping to get beyond Marsh's long reach, Cope avoided the Bridger Basin south of the fort and headed into the little-known Washakie Basin and Ham's Fork areas to the east and north. He told his father hopefully that his diminished outfit was "altogether more agreeable than a military party," but his troubles mounted. Conditions were harsh. Surface water in the arid basin was so bad that the stock almost died from it, and three kinds of gnats, "all of the most bloodthirsty description," tormented the party through the blazing days. Flocks of magpies raided the larder. When Cope arose at five A.M. for a day of fossil-hunting, he found his blankets stiff with frost and the water frozen in the canteens.

Sent after some runaway mules, the teamster spent three days drunk in a dive while the mules starved. Then he stole twenty dollars' worth of supplies in expectation of Cope's firing him, as Cope did, after he "ran the wagon in a ditch and started for another spree." Edward wrote that this abandonment put him "in some bodily peril." When he and his packer went into the steep Wasatch Mountains to study geology, the mules ran away a second time, and the packer went in pursuit, leaving Cope alone with fresh trails of unknown humans. He suspected they'd been made by the Indians or white horse thieves who'd run off his mules, and spent a lonely evening trying to find signs of them in the surrounding hills. Unable to do so, and afraid to camp alone in the area, he rode fifteen miles across the mountains in the dark to get back to his wagons.

By that time, Cope had discovered that Sam Smith's guidance was hardly "first rate." Smith had worked for Marsh the previous year, and, in fact, was continuing to work for him now, as a spy. In July, he'd smugly written to New Haven that Cope "took his meals at Manleys, hi toned for a bone sharp," but had to sleep "in the government hay yard." John Chew, another Marsh employee, wrote to the Professor in mid-August that Smith had quit Cope's outfit and was now with him. On August 28, Smith himself wrote to Marsh, ingratiatingly, "My motive in going with Cope was to ceep him off some places as I think is good bone contry close hear."

All this turned out to be too much for young Edward. His determined letters stopped abruptly in mid-September, and what happened afterward is unclear until, a month later, he wrote to his father: "I am now so convalescent as to be able to write to thee . . . with my own hand. Nothing is left of the fever but the weakness, and I should have been taking the air in short walks, had it not been for other afflictions which have followed the tracks of the first disease." Cope suffered from huge carbuncles on his neck, and "had terrible visions and dreams, and saw multitudes of persons, all speaking ill of me and frustrating my attempts to sleep." Fortunately, he'd made it back to the fort, where his wife came to nurse him and the local community softened its attitude. James Carter's father-in-law was lending him books. Clearly his collecting was over for the year; he departed for the East in late October.

Today's Fossil Butte National Monument northwest of Fort Bridger is typical of the country Cope explored. At an elevation of over 7000 feet, it seemed almost Alaskan when I was there in mid-May. Snowdrifts lingered, stunted aspen groves were leafless, and I could readily imagine what the mosquitoes and blackflies would be like in summer. Yet it wasn't hard to see what had excited Cope. Even in 1997, remnants of prehistory seemed more prevalent than the living ground squirrels, horned larks, and pronghorns. Fossil Butte, a yellow mass of uplifted lake shale, still yields remains of birds, bats, crocodiles, turtles, and fish after a century of intensive collecting. The fish fossils are so abundant that a nearby commercial quarry offers "fossil fishing trips"; participants can excavate up to a dozen complete fossils in a few hours. Nearby, reddish buttes, terrestrial deposits formed beside the lake, still yield bones of mammals like living tapirs and lemurs, along with stranger, long-extinct creatures.

Neither physical nor social obstacles could stop Marsh's rivals from finding fossil vertebrates in such rich environments. Joseph Leidy arrived at the fort two weeks after Cope, but he scooped the younger men by being the first to publish on one of the Bridger field's most spectacular finds. The badlands overwhelmed the middle-aged anatomist, who described them with the romantic imagery of his youth. "No scene ever impressed the writer more strongly," he wrote. "It requires but little stretch of imagination to think oneself in the streets of some vast ruined and deserted city." He confined his bonehunting to short forays around the fort with his friend Carter, and—in the best tradition of the eclectic naturalist—pursued botany and other subjects as well as paleontology. (He also passed the time playing croquet on the fort lawn, as a photo shows.) Yet he was able to dig up a rhino-size, knob-skulled, tusked, herbivorous mammal without living relatives, describe it, and give it the genus name *Uintatherium,* "beast of the Uintahs," in a paper dated August 1. Leidy's description was a little less meticulous than usual, but it was a tribute to his superior knowledge that he was able to beat his ambitious rivals into print despite his limited means.

Cope and Marsh didn't let this inhibit their scramble for priority, and the search reached a fever pitch. On August 17, Cope sent a telegram to the American Philosophical Society naming a similar fossil *Loxolophodon,* which means "crested tooth." As though joining in the spirit of the affair, the telegraph operator changed this to *Lefalophodon,* which means nothing at all. Meanwhile, in New Haven, Marsh was cranking out his own names for *Uintatherium*-like fossils—*Dinoceras* and *Tinoceras.*

Sam Smith's August 28 spy letter to Marsh typified the perfervid atmosphere. "Whe got one tusk and part of the jaw nearly one foot long," he babbled. "I think the same kind that Prof. Lidy got part of the tusk hear that he is blowing about." John Chew, suspicious, wrote to Marsh that some of Leidy's discoveries might have been made "before he was here." A letter from Cope to his father about yet another *Uintatherium*-like fossil, which he named *Eobasileus,* "dawn ruler," verged on the megalo-paranoiac. "In a word, *Eobasileus* is the most extraordinary fossil in North America," he raved. "Marsh and Leidy have obtained it near the same time, and I have no idea whether they have fathered it in advance of me or not." Even the courtly Leidy barely skirted paranoia; he urged James Carter to hide any fossils he came across so that neither Cope nor Marsh would sniff them out.

It would have been confusing enough if Cope and Marsh had confined their Babel to uintatheres, but they feverishly bestowed new names on any bones they could lay hands on that summer. Url Lanham, a naturalist, has written that a genus of early primate which Leidy originally named *Notharctus* acquired six more impressive Greco-Latinate names before Cope and Marsh were through: *Hipposyus, Limnotherium, Telmalestes, Telmatolestes, Thinolestes,* and *Tomitotherium.* None of the six names proved valid, since all applied to the same little creature, which remains *Notharctus,* following the rules of priority. Osborn has joked that a "trinomial system" emerged for naming prehistoric animals: an original Leidy name and the two Cope and Marsh names. (Scientific names are supposed to be "binomial," for a genus and a species, like *Homo sapiens.*) "It has been the painful duty of Professor Scott and myself," Osborn wrote, "to devote thirty of the best years of our lives to trying to straighten out this nomenclatural chaos."

Of course, the paleontologists couldn't be expected to identify immediately the strange animals whose bones they gouged out of the alkaline rocks. Cope surmised from his uintatheres' leg bones that the animals were related to elephants, and he pictured them with trunks. Marsh first thought *his* uintathere was a titanothere like Joseph Leidy's White River Badlands fossil; then he toyed with Cope's elephant idea. Nobody knew in 1872 that elephants and titanotheres didn't evolve until millions of years after the Eocene. Still, the summer's scramble to fling fossil names into print was unseemly, what Url Lanham called "a world-famous debacle in the science of nomenclature."

Joseph Leidy's paleontological career foundered in that debacle, although, since he stayed with the friendly Carters, he had a much pleasanter summer than did Cope. He and his wife liked the West enough to return in later summers, but he gradually withdrew from Western paleontology during the next few years. He said he had trouble getting fossils because Cope and Marsh outspent him, but he once confided to W. B. Scott that he couldn't stand the squabbling. Watching the younger men ignore his prior descriptions and names must have been deeply offensive to the gentle Leidy. "He was as incapable of deceit as he was modest," wrote a friend, "and submitted to imposition rather than enter into controversy." Leidy turned most of his attention to his earlier interest, the study of living microorganisms, about as far from extinct giants as he could get. After his death, in 1891, Osborn discovered there was not a single mention of him in his two rivals' works, "except a brief tribute by Marsh in an early address."

Marsh's activities after he finally went West in the fall are shadowy. The 1872 Fort Bridger army records don't mention him, although they record "Professor Cope's" departure. That year's Yale trip was unpublicized, and only four students participated, though an expedition photo also shows two nattily dressed, unidentified

Professor Marsh (standing, center) and his 1872 expedition

men with a slightly sinister air. "Less is known of the experiences of this expedition than of any of the other three," wrote Schuchert. Most of what is known concerns western Kansas, where Marsh acquired a prime addition to his True West repertoire.

Othniel set the stage with a jaunty self-portrait. "As I glanced at my companions, I was struck by the contrast they presented . . . ," he wrote, "the guide with his wide sombrero hat, his picturesque sash, and his jangling Mexican spurs . . . the Lieutenant, in his army cap, and trim blue uniform . . . The Professor, I regret to say, suffered by contrast with his two companions, as he wore a slouch hat, and an ancient suit of corduroy, with the capacious pockets of the shooting jacket bulging with fossils . . . thus making him the most disreputable looking member of the party." The Professor may have looked disreputable, but he clearly had graduated from ambulance-bound dude to rough-riding frontiersman.

"As we rode up the crest of a high ridge," he continued, "the guide, now slightly ahead, as is usual in Indian country, suddenly called out, 'Great God, look at the buffalo!' and we saw a sight that I shall never forget, and one that no mortal eye will ever see again. The broad valley before us, perhaps six or eight miles wide, was black with buffalo, the herd extending a dozen miles up and down the valley, and quietly grazing, showing that no Indians were near." After "watching the countless throng," they decided to shoot one for supper, and Marsh was chosen to do the deed because his horse, Pawnee, was the best.

Chasing a cow, he found himself in a stampede. "My only chance of escape was evidently to keep moving with the buffalo and press toward the edge of the herd," he wrote, "and thinking thus to cut my way out, I began shooting at the animals nearest to me . . . The whole mighty herd was now at full speed, the earth seemed fairly to shake with the moving mass, which with tongues out, and flaming eyes and nostrils, were hurrying onward pressed by those behind." Not surprisingly, the bison were now more afraid of Marsh and his horse than he was of them, so he was able to force his way behind a butte until the stampede passed. After killing one last cow and taking her hump and tongue, he "sang Pawnee's praises around the camp fire that night, and have told the story of his gallant run many times since."

The dashing frontiersman also made a fossil discovery that neither Cope nor Leidy would ever match. In 1871, he had found the headless skeleton of a six-foot, flightless marine bird contemporary with mosasaurs and pterodactyls; he'd named it *Hesperornis regalis,* "ruling western bird." Now another *Hesperornis* turned up, and enough of its head remained to show that its bill had been lined with teeth. No living bird has teeth, so the evolutionary implications clearly were profound, as Marsh hastened to proclaim in an 1873 paper.

"The fortunate discovery of these interesting fossils," he crowed, "does much to break down the old distinction between Birds and Reptiles, which the *Archaeopteryx* has so materially diminished." Other characteristics, such as a brain cavity much smaller than that of any living bird, supported a reptile-bird link, and paleontologists universally recognized the find as major evidence of an evolutionary transition. "Nothing so startling has been brought to light since," wrote W. B. Scott. Osborn observed that Marsh had "a genius for appreciating what might be called the most important thing in science. He always knew where to explore, where to seek the transition stages, and he never lost the opportunity to point out at the earliest possible moment the most significant fact."

Marsh's triumphs did not alleviate his sense of outrage at Cope's "intrusion." Far from subsiding, the Fort Bridger Babel had barely begun when he returned to Yale in December. First, the two men squabbled over fossil ownership. On January 20, 1873, Cope sent Marsh a letter accompanying some "small specimens I recently received from Kansas as having been abstracted from your boxes!" A collector, perhaps Professor Benjamin Mudge, had sent Cope some fossils already promised to Marsh. The letter was polite enough, and Cope called Marsh's toothed birds "simply delightful," but the envelope was addressed to "*O See* Marsh," a mocking *sobriquet* coined by New Haven's young ladies in response to Othniel's pomposity. Marsh's January 27 reply was brusque. He thanked Edward ironically for the "abstracted" fossils and continued, "Where are the rest? And how about those from Wyoming?" With solemn duplicity, he rebuked Cope for employing his spy, Sam Smith, "to whom I had given valuable notes about localities." And he threatened, "I came very near publishing this with some of your other transgressions . . . I was never so angry in my life."

An aroused Cope shot back with counteraccusations—"some appropriative person" had stolen one of *his* fossils—and grandiose assertions. "All the specimens you obtained during August 1872 you owe to me," he wrote. "Had I chosen, they would have been all mine. I allowed your men Chew and Smith to accompany me and at last when they turned back discouraged, I discovered a new basin of fossils, showed it to them, and allowed them to camp and collect with me for a considerable time."

Marsh responded that he desired "most sincerely to be on

friendly terms" but that Cope had "deeply wronged" him many times. Not only had Cope suborned Smith; he had snooped around his specimens in Wyoming and perhaps stolen some. Marsh also suspected him of retaining some Kansas fossils. "Now for all this, you have only yourself to blame . . . ," Othniel wrote. "Could I have done less than to give you a chance to explain the matter? You have said distinctly that you have neither Wyoming or Kansas fossils of mine, and I have, therefore, nothing more to say." Soon after this exchange, the two met in New York at a party held in honor of the English scientist John Tyndall. Cope told his father that Marsh "stuck to me like a leech and I hope became fully satisfied that I was not a thief. It seems persons had been writing to him and had wronged me very greatly. As to the dates I said nothing."

Those "dates" caused even more trouble. Cope's hastily published articles on uintatheres were such a tangle of confused and contradictory statements that Marsh decided Cope had been falsifying publication dates to beat him into print. Eight Cope papers in the *Proceedings of the American Philosophical Society* were dated August 15, 1872, but Marsh pointed out that papers were supposed to be dated for the meeting at which they'd been read, and that there had been no meeting of the society on August 15. Marsh concluded that the papers had not been published until October 29, and he tried to get the Philosophical Society to censure Cope. A vote to do so was defeated, but the society's secretaries evidently sat on the younger paleontologist in other ways. "I have another lesson of the weakness and depravity of human nature on hand," Edward wrote to his father, "which confirms previous ones."

Cope and Marsh then began trading insults in the sedate *American Naturalist.* A Cope article in March described three species of his uintathere genus *Eobasileus* and predicted that they would "be found to be the predecessor in time of the huge probiscideans now known." Marsh, having concluded that uintatheres were unrelated to elephants, said so abrasively in the same issue. "Prof. Cope . . . has made several serious mistakes in his observations," he gloated. "He has likewise been especially unfortunate in attributing

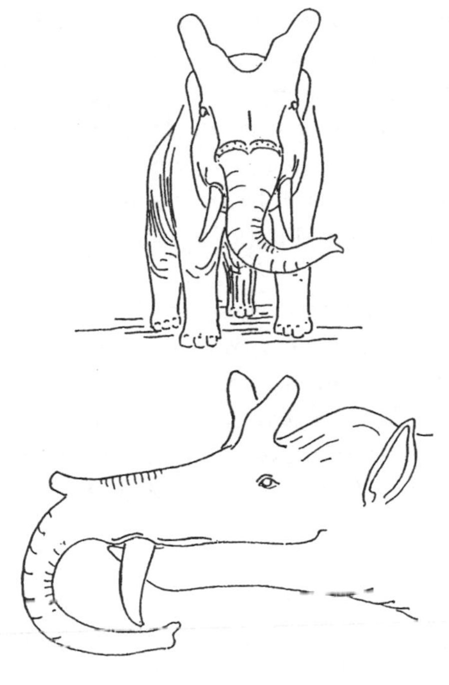

Cope's sketches of a uintathere

to the Dinocerata characters they do not possess; and hence his conclusion that all these animals are . . . possessed of a proboscis is quite erroneous. In his references and dates Cope also has shown the same inaccuracy that marred his scientific work." Marsh kept up his gloating in the April issue, in which he claimed that Cope's *Eobasileus* article had contained "no new facts on the subject but some interesting additions to the list of errors I have pointed out in the same number," and that *Eobasileus* actually belonged to *his* genus, *Tinoceras*.

In May, Cope insisted that uintatheres did indeed have trunks, and jeered at Marsh for going on "as though Uintatherium were a Rosinante and the ninth commandment a wind-mill." He asserted that he couldn't be held responsible for the errors of printers and society secretaries while he was out West, and he included a signed statement from the printers, acknowledging their errors. "To sum up the matter," he wrote, "it is plain that most of Prof. Marsh's criticisms are misrepresentations, his systematic innovations are untenable, and his statements as to the dates of my papers are either criminally ambiguous or untrue." The May issue also contained Othniel's complaints about Edward's refusal to share his Fort Bridger papers. "I received nothing of the kind from Professor Cope," he wrote aggrievedly.

At this point, the *American Naturalist* editors relegated the squabblers to appendices "at the expense of the author," as though they were sending bad boys to the cloakroom. Marsh had the June appendix to himself, and carried on for nine pages. "As a sleight-of-hand performance with names and dates, it shows practice and is amusing," he wrote of Cope's explanation about printers' errors, "but to those familiar with the subject, and to moralists, it suggests sad reflections." He accused Cope of having written the printers' statement himself. "For this kind of sharp practice in science, Prof. Cope is almost as well known as he is for the number and magnitude of his blunders," Marsh said with disdain. ". . . Prof. Cope's errors will continue to invite correction, but these, like his blunders, are hydra-headed, and life is really too short to spend valuable time in such an ungracious task."

Cope had the last word, such as it was, in a one-paragraph appendix in the July issue: "The recklessness of assertion, the erroneousness of statement, and the incapacity of comprehending our relative positions, on the part of Professor Marsh, render further discussion of the trivial matters on which we disagree unnecessary; and my time is too fully occupied in more important subjects to permit me to waste it upon personal affairs which are sufficiently before the public."

It is a tribute to Cope's hyperthyroid vitality that he didn't wilt before Marsh's caustic, partly accurate attack. "As to the learned professor of Copeology in Yale," he wrote to his father in December 1873, "he does not disturb me, and as I promised in my last reply (last spring) I will not notice him again. The longer he continues on his course, the more injury he inflicts on himself . . . I will however correct his errors when they come in my way, as is the custom of naturalists, and I expect the same treatment. This need not and should not excite any personal feelings in any person normally or properly constituted; which unfortunately Marsh is not. Unfortunately, I say, because he makes so many errors and is so deficient that he will always be liable to excitement and tribulation. I suspect that a Hospital will yet receive him." Brave words, but it was Edward who had emerged feet-first from the badlands at Fort Bridger.

7

Marsh the Reformer

Marsh's public response to vexatious 1872 was to act as though it hadn't happened. His 1873 Yale expedition reprised the 1870 and 1871 trips, including eleven students as well as two scouts and Oscar Harger, the assistant. They dug for Miocene camels and horses along the Niobrara River, returned to Fort Bridger for more uintatheres, and gathered a record forty-nine fossil boxes. Again, a student publicized the adventure. "Nearly every day of our outward march," wrote T. Mitchell Prudden in a letter to the *New York Tribune,* "the thermometer indicated from 98 to 104 degrees in the shade." Again they rested in Salt Lake City, where Brigham Young questioned Marsh about his horse fossils, explaining that the Latter-day Saints' foes had used mention of prehistoric North American horses in the Book of Mormon as evidence that it was not divine revelation. "So it seems that while most theologians were regarding the developments of science with fear and trembling," said another *Tribune* letter, "the chiefs of the Mormon religion are prepared to hail the discoveries of paleontology as an aid in establishing their peculiar beliefs."

Yet the 1873 expedition was noticeably less rah-rah than the earlier ones. The student Henry Farnam, later an economics professor at Yale, described it as mostly drudgery; he complained that Marsh explained very little about what they were doing, perhaps because he was "afraid that if he told us anything, it might . . . leak back to his antagonist." Returning east early, Farnam was startled when he was asked about an "exciting bear hunt." After replying that "we never saw bear or anything on the trip that was the least exciting," he was shown a newspaper clipping that was "an account of a fight which the Marsh party had with a grizzly bear, and in the end the bear was killed by one of the group, who thrust a big hunting knife into his heart." Farnam, recalling that one student had carried such a knife, realized "that the whole story had been concocted by a few of the group who had felt very much disappointed that they had had no thrills on the trip and, therefore, resorted to fiction."

Marsh's image was wearing thin in spots. That year, *Buffalo Land,* a pastiche of tall tales and immigration come-ons by a Kansas booster named William E. Webb, lampooned a "Professor Paleozoic" who led student expeditions. Slyly evoking the Onondaga Giant affair, Webb wrote that "when the Professor said a particular rock belonged to the cretaceous formation, one might safely conclude that no modern influences had been at work." He also had his myopic caricature mistake surveyors' signs for aboriginal rock carvings and Indian hatchet marks for prehistoric bird tracks. "He could tell cheese from chalk under the microscope," Webb allowed, and—no stickler for consistency—made his victim parrot the claims of anti-evolutionist boosters. Finding human footprints in Cretaceous rocks, Professor Paleozoic cried, "What folly to suppose that such a land, so peculiarly fitted for man's enjoyment, should remain, through a long period of time, tenanted simply by brutes," and prophesied that the Great Plains would become the world's most abundant wheat-producing region before 1890.

Webb then added insult to injury by filling two chapters with admiring quotes from "the eminent naturalist Prof. Edward D. Cope, A.M.," who had "visited the plains, and spent some time in careful exploration." Cope's imaginings about life in Cretaceous seas evidently appealed even to the anti-evolutionists, perhaps because the strange, vanished sea serpents seemed less Darwinian than Marsh's toothed birds, or simply because Cope described them more vividly than Marsh. For someone who had tried as hard as Othniel to cut a popular figure out West, this would have been humiliating.

It was the end of an era for Marsh. Students' collections no longer satisfied him, and he never even opened many of the expeditions' boxes. He needed a more businesslike approach. Charles Schuchert observed that he had two main purposes: to pre-empt every promising new fossil field and to get all the fossils possible of the animals he wanted to write about. "To develop these two lines of field work," wrote Schuchert, "he turned to paid collectors."

Marsh's collectors fell into two categories: riffraff like Sam Smith and educated types like Benjamin Mudge, whom Cope had found "left behind" at Fort Wallace. Mudge had been a natural history professor and a geologist before the West's boom-and-bust financial turmoil left him unemployed. Marsh treated thugs and former professors about equally, however; he expected them to collect solely for him and to conceal their finds from rival collectors. Collaboration was out of the question. This caused resentment, as did his frequent failure to pay men on schedule, although that arose more from absentmindedness than stinginess. Despite the discontent, Marsh's collectors had grown to at least a dozen teams by 1874, when he began to confine his Western travel to quarry-checking jaunts. But there was to be one more big adventure.

Othniel didn't plan to travel at all in 1874, because construction of the Peabody Museum was under way. The West became more dangerous than ever that July, when Colonel Custer led a thousand-man expedition into the Black Hills. The region of pine-forested granite just west of the White River Badlands is the center of the Lakota universe, and the local Oglala chief, Red Cloud, and other tribal leaders had been very clear about including it in the "permanent reservation" they had negotiated in 1868. The government nevertheless decided to build the Great Northern Railroad across the reservation, and the army thought a Black Hills fort would be advantageous. Scouting for a fort location was the official reason for the illegal invasion—but the hills were rumored to have gold, and Custer confirmed this. "I have upon my table 40 to 50 particles of pure gold," he wrote in an August 2 dispatch; "veins of what the geologists term gold-bearing quartz crop out on almost every hillside."

The army had invited Marsh to join the Custer expedition, but he had declined and sent two graduate students instead. As his artifact collections suggest, he had more sympathy for Native American cultures than did many of his contemporaries, although his refusal probably arose more from circumstances than from his respect for sacred land. The Professor was no cultural relativist. When Pawnee scouts refused for religious reasons to hunt for bones, he showed them the likeness between a fossil horse jaw and the jaws of their mounts, and that allegedly turned the scouts into bonehunters. On a less uplifting note, he robbed Indian graves, telling squeamish students that he needed skulls to study "the origin of the Indian race"—not that he did so. Evidently unaware that the Lakotas originally had been town-dwelling farmers, he said at an 1875 hearing that Indians "could be best managed" by keeping them on the reservations, because "many years, if not decades, would pass before their nomadic blood could be tamed to ordinary agriculture."

Yet when the army invited him again in October, this time to dig in the White River Badlands, Marsh decided to go, after all. He may have had more than fossil-hunting in mind; he previously had ignored the area because Hayden and others had picked it over. His correspondence shows that he'd been studying U.S. Indian policies critically since the early 1870s. By now, Marsh's government contacts had made him something of a political insider, and he may have seen value for his professional status in those policies. He doubtless felt real concern for Indian rights, as well as a recognition of the potential for publicity in advocating them. Indian agency corruption was a hot topic, mentioned prominently in Mark Twain's and Charles Dudley Warner's eponymous 1873 satire, *The Gilded Age*. Even the *Herald* deplored it, likening it to stealing from orphans.

The turbulent White River Agency, where Red Cloud and his

large Oglala band wintered, was the heart of the conflict over corruption. The 1868 treaty had promised to supply the Oglalas with food and clothing in exchange for land concessions, but the arrangement turned into a source of constant dissatisfaction, often of real deprivation, and the Black Hills invasion had exacerbated the distrust and anger. Red Cloud, among the fiercest war leaders of the 1850s and 1860s, was also one of the smartest, and he saw that fighting would not stop the deprivations and invasions. Looking for political leverage, he apparently was quick to see the potential of an Eastern celebrity's sudden descent on his tribe, even one as bizarre as a bonehunter.

The *New York Tribune* covered the badlands trip in a lengthy story on December 22, 1874; it must have come from Marsh, since it was datelined New Haven. It reported that the Professor had decided to go because recent fossil finds there threw "additional light on the problem which he was then engaged on working out," although, typically, he didn't identify the problem. The time was "unpropitious" for a visit to the agency. A fight almost had ensued after whites raised an American flag and Indians cut it down. When Marsh arrived with a military escort from Fort Laramie on November 4, Agent J. J. Saville was withholding fall rations until the Indians agreed to cooperate in a census, which many refused to do, particularly several thousand "Northern Indians, of the wilder tribes," who had come to the agency for the winter. "Outlaws, renegades, and 'bad Indians' swelled the numbers that surrounded the Agency, and made the neighborhood unquiet, not to say dangerous," the *Tribune* reported. Saville, after advising Marsh to get an escort from the Oglalas, called together their leaders to propose this.

"As soon as they were brought together," the *Tribune* continued, "it became evident that they mistrusted the intentions of the bone hunters . . . White Tail, one of the principal chiefs, sprang at once to his feet and harangued the audience, recounting previous grievances and declared that the proposed bone-seeking was merely a ruse to begin digging for gold and invading the Black Hills region. His speech evidently conveyed the sentiments of the other chiefs; they listened intently, giving vent to applause and sympathy with guttural ejaculations of 'How! How!' But a speech from Prof. Marsh through the medium of an interpreter, promising that their complaints should be heard in Washington, stating specifically the objects of the expedition, and holding out the prospect of pay for Indian services in bone-hunting, turned the scales at once." The Oglalas agreed to let the expedition continue, on condition that "a selected guard of young warriors" accompany it, ostensibly to protect it against Northern Indians camped across the White River—but actually to spy on it.

A snowstorm the next day delayed Marsh's start, at which point Agent Saville suddenly released the Oglalas' rations, which "quite changed the aspect of affairs. Having got their annuities, the Indians were no longer on their good behavior . . . The bone-hunting expedition was discussed in every lodge. They all arrived at one conclusion, that the pretense of seeking fossils was much 'too thin.'" When Marsh and his party arrived at the agency to pick up their escort two days later, they encountered a hostile crowd "armed quite as well as the soldiers, with breech-loading rifles and revolvers of the most recent pattern." The escort refused to accompany the expedition, and a man named Pretty Crow "suddenly precipitated a crisis by shouting, 'the white men are going into our country to find gold: we must stop them at once.'" Outnumbered thirty to one, Marsh and the soldiers retreated a mile and a half to the army post at Fort Robinson while the crowd "showered insults" on them. "It is not worthwhile to repeat here the jeers which this movement elicited from the Indians," the *Tribune* added primly. "The language of signs is never more efficient than for such a purpose, and it was freely used."

Someone then convinced Marsh that he might win the Indians over if he gave a banquet for fifty of their leaders. "The feast was given in a tent 30 feet in height, and every detail of Indian etiquette was strictly observed. At its close . . . a reluctant consent was again accorded, with the warning that the Miniconjous were likely to kill the Professor if he crossed the White River." Again the Indians promised Marsh a band of scouts, under the leadership of Red Cloud's son-in-law Sword, but this time Marsh was skeptical. He "sent word quietly, late in the night after the feast, to his interpreters and guides, to be ready the next morning. The dread of the Minneconjous and Uncpapas overcame the blandishments of the feast. Indian scouts, guides, and interpreters all alike refused to go."

The *Tribune* continued, "Disappointed and not a little exasperated by these repeated delays, Prof. Marsh resolved on the most extraordinary move of the expedition. He decided to give the Indians the slip . . . Shortly after midnight . . . marching down between the Indian villages as silently as possible, the expedition sought the White River at the only spot where, for at least 15 miles, it was fordable. The dogs barked furiously as the party defiled between the lodges, but fortunately their owners slept." At daybreak, Lakota sentinels appeared on the buttes, and the expedition quick-marched, because of the cold and anxiety, but they made it to Marsh's fossil site.

The bones were spread over ten miles in a deep gorge, and the party got to work immediately. It was too cold to do anything but work. "They could not sit on the ground to dig; moving about was necessary to keep from freezing," reported the *Tribune*. A glass of water had to be drunk in a gulp before it froze, and Marsh broke icicles off his mustache so that he could eat. When the weather warmed, snow fell so fast that fossils had to be swept with makeshift brooms to prevent their disappearing. Indians came into the camp by daylight to spy, and one night they "attempted to surprise" the men, but the guards were alert, and "the Indians, perceiving that the camp was alarmed, withdrew."

The Professor found some time for "Copeology" even in this tight spot. "One morning Marsh and I went from camp on our usual exploring trip for fossils," recalled the 9th Infantry's Major Andrew S. Burt. "We hadn't gone far when, passing by a high-cut bank, I happened to glimpse a fossil bone sticking out of that bank. I pointed it out to Marsh. The instant his eye caught the object he wheeled his horse, dismounted, and rushed to what afterward looked to me like the leg bone of an ox. Marsh dug around the specimen awhile, then suddenly seemed to have gone crazy. He danced, swung his hat in the air, and yelled, 'I've got him, I've got him.'" According to Burt, Marsh was "having an exciting controversy with a rival scientist about a prehistoric animal which both claimed to have discovered and named."

The expedition unearthed over two tons of fossils, "rare specimens, illustrative of entire classes of quadrupeds, of which all that is known has been ascertained within a very few years." Despite the cold and danger, Marsh persisted in removing the bones "entirely" from the deep gorge. Another snowstorm threatened as the men began to pack, and "a more serious cause for alarm was found in the representations of a party of Indians, led by Spider [Red Cloud's brother] and Sword, who came to warn the expedition of its immediate dangers. They had ascertained that the Northern Indians were taking their wives and children to the Black Hills, preparatory to making an attack on the camp. There was good reason to expect the attack that night. To throw the specimens in the wagons and rattle off with them unpacked was simply to break them to pieces. To pack them at night, burning lights in the tents, would be to invite attack. The Indian asks no better mark for a shot at long range than a lighted tent. Great was the risk of remaining." It was a cliff-hanger, but paleontology prevailed. "Prof. Marsh, after due consultation with the officers of the escort, decided to stay long enough to pack properly. The expedition broke camp the next day, and not too soon; subsequent reports state that a large war party of Northern Indians scoured the Bad Lands on the following day."

One of Marsh's assistants later remarked that the Professor was never in any greater danger from Indians than when he entertained them in New Haven, and the malevolent Miniconjous do seem shadowy, along with some other aspects of the expedition. The Oglalas' obliviousness of Marsh's midnight dash into the badlands is surprising, considering their previous resistance to his entry, and his fossil site has faded into obscurity. When I inquired about this at Badlands National Park, the paleontologist Rachel

Benton doubted that anyone knew where the "deep gorge" was. She suggested I ask at Agate Fossil Beds National Monument, because its original ranch owner, James H. Cook, was said to have known Marsh, but I had done that just the day before, with no success.

Agate, an afternoon's drive south of Badlands, illuminates Miocene prehistory vividly with the diorama of dinohyids and daphoenedons, but its historical past is murkier. James Cook said that it was he who had been Marsh's interpreter to Red Cloud. "I heard a great deal of the Indian side regarding bone-hunting in the Sioux hunting grounds," he wrote in his memoirs. "I met Professor Marsh and talked to him. Then I went to Red Cloud's lodge and talked the matter over with him. I told him that Professor Marsh was a friend of the Great Father at Washington; that, if he was allowed to hunt for stone bones, I thought he would be a good friend of the Sioux people." Cook was a teenager in 1874, however, and when a Marsh biographer queried him in 1931, he replied, "Gladly would I aid you in the work you mention, but I fear that I can no longer put much confidence in my memory regarding Professor Marsh."

Still, there's no real reason to doubt all of Marsh's rousing tale. Its obscurities were probably further manifestations of his growing secretiveness. He did collect in the White River Badlands in 1874; the fossils are in the Peabody Museum. The Lakotas were in a dangerous state of unrest. He did go on to become the period's most famous advocate of Indian rights, though some aspects of that advocacy are unclear.

The *Tribune* story said nothing further about it, only that Red Cloud had invited Marsh to a "dinner of dog" on his return to the agency, which he declined more from fear of having to reciprocate with another costly banquet than from distaste for "fricasseed puppy." Later accounts say that at some time before or after the bone hunt, Red Cloud gave the Professor samples of bad treaty rations—moldy flour, maggoty pork, spoiled sugar, coffee, and tobacco—and asked him to take them to Washington as evidence of Agent J. J. Saville's poor management. Marsh promised to do so. "Then the Professor and the Indian Chief parted," wrote Charles

Schuchert, "the latter doubtless with little faith that this white man would keep his promises any better than the others with whom he had come into contact."

Marsh did keep his promises, even though it took him five months to do so. Recently elected to the National Academy of Sciences (with one dissenting vote: Cope's), he took Red Cloud's samples to Washington when he attended academy meetings there in April, and showed them to E. P. Smith, commissioner of Indian Affairs. Smith's boss, Interior Secretary Columbus Delano, was one of the crookedest members of the notoriously corrupt Grant cabinet, so the commissioner showed scant interest in the unpleasant objects. Marsh was not discouraged, and his flair for publicity came to hand. Escorted by the postmaster general, a Connecticut acquaintance, Othniel took his story right up to President Grant. As the *Tribune* article shows, he also had pull with the high-minded, antiadministration press, and his efforts received conspicuous coverage.

Unable to ignore him, the Bureau of Indian Commissioners deigned to meet him in New York on April 28. There, Marsh showed them his samples and diplomatically noted that their agents and traders, though not really dishonest, were incompetent to handle the large volume of goods and money passing through their hands. When asked his "view of the Indian question generally," Marsh replied, according to an April 30 *Tribune* article, "that he had little hope for the Indians till the whole present system could be changed, especially the issue of food and goods." Marsh cited the issue of annuities to the thirteen thousand Indians at the Red Cloud agency, which had been "such a hurried affair that no one could really know whether half the goods intended for the Indians were delivered or not. After the issue, there was great discontent among the Indians."

Representative W. R. Steel of Wyoming denied the charges of incompetence and corruption, but it was too late to muffle Marsh. Spurred by his testimony, the *Tribune* began an editorial campaign to clean up the Interior Department. When the publicity drove Secretary Delano on May 10 to propose a commission to "investigate

Interior Secretary Delano raging at Marsh

certain reports put in circulation by a Mr. Marsh, relative to the Indian Service at Red Cloud agency," Othniel's years of networking bore fruit. "The *New York Evening Post,*" wrote Schuchert, "rebuked Secretary Delano for venting his spite by so childish a trick as referring to a man of Professor Marsh's standing as 'a Mr. Marsh.' The *Tribune* practically burst into flames at the insult; editorial writers and reporters had a field day; and thousands who knew little about Indian frauds, and cared less, laughed at the controversy." A May 29 cartoon in *Frank Leslie's Illustrated Weekly* showed a lilliputian Delano raging on the boot tip of a comfortably seated professor who was obliviously reading the *Tribune.* "Columbus Discovers 'A Mr. Marsh,'" jeered the caption. The *Tribune* began to call for Delano's resignation.

Then Red Cloud almost ruined things. Summoned to Washington in May to air his complaints and discuss the Black Hills, he was strangely evasive when he met Marsh, the assistant interior secretary, and Commissioner Smith. Apparently, the spoiled rations were not as typical as the Oglalas had led Marsh to believe and may have been chosen for their nastiness. "This failure of Red Cloud to back up his statements left Professor Marsh in a very embarrassing position," wrote Schuchert, "even when Red Cloud later blamed some of his evasiveness on the interpreters, whom he claimed not to have understood."

Marsh's discomfiture caught the attention of the *New York Herald*, which so far had ignored the affair, although it had been covering the Black Hills gold story at length. (One issue carried both an editorial admonishing would-be miners to keep out of the region and a large map with detailed directions for getting there.) The Professor's lobbying for Indians would not have impressed the paper's owner any more than would his hobnobbing with Mormons (whom Bennett called "a blot on the republic," perhaps in envy of their polygamy). The account of Marsh's exciting badlands adventure in the December *Tribune* certainly would have displeased jealous Jimmy. Why couldn't the Professor have narrowly escaped the malevolent Miniconjous while the *Herald* was covering him? Bennett's eyes would have narrowed at the thought of the midnight dash past the sleeping Oglalas. Was the giant debunker getting into the news-fabricating business?

On May 15, the *Herald* ran a short piece, "Professor Marsh Contradicted by His Interpreters," about an Indian translator who denied that Red Cloud had complained about the rations to Marsh at the agency. It paid more attention to Red Cloud's visit after that, with sardonic coverage of his party's complaints about hotels and white duplicity. "Red Cloud should be informed at once by Professor Marsh or some other trusted friend," quipped a May 20 piece, "that there is but one place on this continent where they can hear the strict truth"—Tammany Hall. A May 22 article defended Secretary Delano from an attack by an Indian named Lone Horn. And a week later, the *Herald* said of Marsh that "the case he had heretofore presented and into which he was led by Red Cloud was not sustained."

Red Cloud and the other Lakota leaders returned west after refusing a princely $25,000 for the Black Hills, however, and Marsh could resume his campaign without embarrassment. In June, Agent J. J. Saville demanded an investigation of Marsh's charges, and Secretary Delano appointed a committee, the first step toward strangling the affair with red tape. The Professor was not to be gagged, however. "Possibly his days with the Army had taught him the value of a surprise attack," observed Schuchert. "Most certainly, he was aware of the value of wide publicity."

On July 10, Marsh sent President Grant and his cabinet a letter calling Saville "wholly unfitted for his position and guilty of gross frauds upon the Indians in his charge." The letter repeated his other accusations and added that the Indians actually had been starving and freezing during the past winter because of fraud and mismanagement. It closed with a stirring (if disingenuous, considering that Grant had vetoed a bison-protection bill) appeal to the President: "You alone have the will and power to destroy that combination of bad men, known as the Indian ring, who are debasing this service and thwarting the efforts of all who endeavor to bring to full consummation your noble policy of peace." Marsh's thirty-six-page pamphlet detailing the Red Cloud situation accompanied the letter. He released both to the press the next day, and sent copies to fifteen hundred prominent people a few days later.

Administration newspapers fought back, accusing Marsh of fomenting the controversy for personal gain and publicity, of holding "sculking [sic] interviews" with Red Cloud, and of being "impetuous but not very practical—a nervous, impulsive, credulous man, naturally combative." But public opinion had turned against the administration. Without getting elected or appointed to any office, Marsh had imposed his carefully crafted image on the highest level of national policy, a rare achievement. "There is something fascinating to the average intellect in the way in which the climax has been reached," observed an editorial in the Reverend Henry Ward Beecher's *Christian Union.* "The Professor with his party of retainers and a small cavalry escort, trying to penetrate the remotest wilds of our territory, and caring little for the red man unless he might chance to be found in a fossil state; the encounter with the unwilling and unscientific savage in a state of activity not in the least suggestive of fossils; the tedious 'talk' about the council fire; the final conviction of the Professor that in order to get at his beloved hipparions and pterodactyls he must covenant with the legitimate owners thereof; then the circumlocutions and evasions of the Washington officials, and finally the formal appeal to the President—all these combine to make a passage of departmental history that shall be a warning to future Secretaries of the Interior and Indian Commissioners."

A measure of Marsh's success was that the *Herald* changed its position, rather as it had retreated from the Onondaga Giant. This meant that James Gordon Bennett, Jr., had changed *his* position; *Herald* policy always came from the top. "He required that all things should be made known to him before they were attempted," observed Don Carlos Seitz, editor of the *World,* "and the office council, playing safety as much as possible, laid everything in his lap." No initiative went unpunished. When his editors once told him that a man was indispensable, Bennett demanded a list of indispensable staff members and fired them all.

In a May 22 *Herald* interview, none other than Colonel Custer called Marsh "a disinterested witness," which evidently gave Bennett and his zombies pause, because the May 24 issue called for an investigation of Indian Service corruption. Yet the *Herald* waited until Delano's goose was well cooked before it joined the feast. It wasn't until July 21 that an article, "The Indian Ring," on Marsh's evidence to the Investigating Committee, conceded that "the case against Saville at Red Cloud is pretty clear." Another article, two days later, "The Indian Fraud," implicated Delano and called Marsh's report "truthful," but the *Herald* still didn't climb on the professorial bandwagon. An August 15 piece attributed the public's awareness of the Indian fraud to "the columns of the *Herald,*" without mentioning Marsh, although the next day's edition did note in passing that the Interior Department's culprits' "paths to the outer world and away from the Department are being made by the *Herald,* Mr. Welsh, and Professor Marsh."

Secretary Delano kept flapping around Washington, but the smell of roast goose had attracted more reformers, including the *Herald*'s "Mr. Welsh," a Philadelphia businessman who had been trying to change Indian policy since Grant's first election, and a disgruntled and well-informed former Indian Board clerk. The investigative committee went to Cheyenne, where it listened to "more lying in this town . . . than there has been since it was settled," as a Wyoming hotelier told the *Tribune.* The attacks on Marsh grew nastier. He was accused of unmentionable frontier behavior, and, in the *Springfield Republican,* of being "a depraved member of the Christian religion." Delano even reviled Marsh in person when both happened to have breakfast at Wormley's Hotel on September 10. According to *The Nation,* Delano called his tormenter a poltroon and a liar as Marsh calmly took notes. Delano's son and the waiters, however, were the only witnesses.

The *Herald* chose September 10 finally to grant Marsh celebrity status; an editorial that day characterized him as being of "high personal character" and urged Grant to dump Delano. "Professor Marsh's catalogue of lies emanating from the Indian Bureau and the Interior Department is a curious study," the editorial said. "No officers were ever before so fearfully exposed as are Commissioner Smith and Secretary Delano by the earnest and empathic Yale Professor. One falsehood after another is fixed upon . . . Professor Marsh is to be commended for his disinterestedness and courage in this matter of the Indian frauds, especially as it will be impossible for the Commission to investigate his charges and meet them successfully with a whitewashing report after such plain speaking."

Delano didn't last as long as it took the commission to administer a 929-page slap to the department's wrist in mid-October. The *Herald* printed his resignation letter on September 27 under another long "Indian Fraud" article. "No correspondence between two U.S. officials could give as much gratification to the country," it crowed, adding that the resignation had been "for years the wish of the country." Marsh could have pasted the article in his scrapbook with satisfaction, although the *Herald* didn't gloat over Delano's fall as lovingly as did papers like the *Washington Capital,* which, on October 9, ran a bit of doggerel by Samuel Walker, the disgruntled Indian Board clerk:

R.I.P.

Not a drum was heard, not a funeral note,
As the corpse to Ohio was hurried,
Not an injun discharged a final shot,
O'er the grave where Columbus lies buried.

Some reflection on James Gordon Bennett's character, of course, might have tempered Marsh's satisfaction. It was the second time the Professor had helped the *Herald* to do an about-face, a maneuver detested by its owner, who liked to be called the Commodore, because he'd briefly commanded his yacht for the Union during the war. (He had conceived his owl fetish when sylvan hooting saved him from running aground in a fog.) Yet Marsh doubtless gave little thought to Jimmy's feelings. He had plenty of newspapermen on his side.

The departure of Delano and Saville also must have gratified Red Cloud and his people, although it's unclear what the Oglalas gained from Marsh's political *tour de force.* Red Cloud's biographer Robert Larson wrote that "no important reform of the reservation system occurred because of it." That was hardly Marsh's fault, given the Grant administration's corruption, and it didn't stop Red Cloud from sending him a peace pipe through Lieutenant W. L. Carpenter, an 1874 expedition member. (Carpenter had caught bonehunting fever in the badlands and during later hostilities proposed to send Marsh a Lakota warrior whom his Crow scouts had chopped into six "convenient" pieces.) The lieutenant wrote to New Haven from Fort Robinson in 1877 that "Mr. R.C. came to my quarters and said he wanted to give it to you, and wished me, at the same time, to tell you something. This is his speech, as translated by an interpreter."

Marsh and Red Cloud in 1883

> I remember the wise chief. He came here and I asked him to tell the great Father something. He promised to do so, and I thought he would be like all white men, and forget me when he went away. But he did not. He told the great Father everything, just as he promised he would, and I think he is the best white man I ever saw. I like him.

Red Cloud visited Marsh in 1883, and they were photographed together holding the peace pipe. The *New Haven Register* reported that he gave "not even a grunt of appreciation" at the Peabody Museum but displayed "a grim smile of pleasure" at the Winchester rifle factory, although nobody seems to have known enough Lakota to get a more informative response from him.

On the other hand, Marsh gained a clear advantage from his new prominence: stronger influence in the growing government science establishment. Exactly how he came to wield this influence is less clear. He somehow let a veil fall over his less altruistic activities. Nonetheless, it's possible to trace his overall progress toward paleontological supremacy. The key to the power he wanted lay with geological surveys like Ferdinand Hayden's, although Hayden himself was not an ally. Their relationship had paralleled the Marsh-Cope one, degenerating by 1869 from cordial fossil-sharing to Marsh's accusations that Hayden was encroaching on his territory and appropriating his specimens.

Marsh had been chagrined when Hayden lent his facilities to Cope, however parsimoniously, and angered when he let Cope use survey bulletins to taunt his Yale rival. Cope had incited Hayden to start those bulletins in 1873 "to break the monopoly on scientific publication held by the *American Journal of Science* and its backers in eastern academic circles." Again, Marsh was furious in 1874 when Hayden claimed that his name was "being used extensively here at this time by certain parties to sanction a statement that the survey of which I have charge is a fraud"; he saw that claim as a threat to blackball his impending election to the National Academy of Sciences.

Othniel's near-unanimous election to the academy, however, demonstrated a shakiness in Hayden's reputation. Despite his popularity in the West and with the *Herald,* many Eastern political and scientific insiders despised the survey leader. His biographer Mike Foster called them "a hidebound portion of the intellectual community which disdained popularizing science at which Hayden was an acknowledged master," but conceded that his subject annoyed almost everyone who knew him, at one time or another, and thrived on controversy. By 1874, Hayden had alienated such powerful figures as John S. Newberry, a geologist with whom he'd worked in the early 1850s, and the New York geologist James Hall. And there were surveys other than Hayden's. His own popularity had helped them expand.

Clarence King, the boy wonder of American geology, ran a major survey along the 40th parallel, the Union Pacific Railroad's chosen route. King had performed prodigies of Western exploration in his early twenties; he was only twenty-six when he received congressional funding for his survey in 1867. Like Marsh, he espoused the new specialized approach to science, and his specialty was economic geology, the study of ore and coal deposits. He also gained a reputation as a scientific sleuth when, in 1872, he exposed a Colorado scheme to salt the ground with gemstones and sell millions of dollars' worth of bogus mineral claims. That feat had saved many investors from bankruptcy and made him popular in the world of finance. A close friend of the patricians Henry Adams and John Hay, he seemed bound for glory in 1875. Then, too, he happened to have been a classmate of Othniel's at the Sheffield School. Marsh had been publishing in King's survey bulletins for several years by 1875, including reports about his major work on horse evolution.

An even more promising connection was Major John Wesley Powell, head of a Colorado River region survey since 1868. Powell lacked King's and Marsh's access to big capital, but he compensated for that with drive and political adroitness. Toughened by a hardscrabble Midwestern boyhood, he'd picked up his scientific training while teaching at rural schools. He had fought in the Civil War and lost an arm at Shiloh. In 1865, he was appointed professor of geology at Ohio Wesleyan, and had anticipated Marsh by leading the first student expedition to the West in 1867. Two years later, his fame approached that of Lewis and Clark because of his epic navigation of the Colorado River through the unexplored Grand Canyon. "He had unlocked the last great unknown region of the country and made it his own," wrote Wallace Stegner. "By the end of his career, he would know the West as few men did, and understand its problems better than any."

Powell's photographs show a disheveled version of Marsh's glowering determination, and it would have gone hard for the Yale Professor if Powell had been a paleontologist. The major had begun his career with the broad-brush naturalist's enthusiasm that led some to bonehunting. But by 1875 his interests had narrowed to

Western settlement and ethnology. He disagreed bitterly with the optimistic Hayden faction about settlement, advocating limited development based on irrigation rather than headlong expansion in hope of increased rainfall. He disagreed just as bitterly with Western boosters about Indian policy, pressing for education and civil cooperation instead of the prevalent military repression. Marsh's squabbles with Hayden and Interior Secretary Delano must have tickled him, so he was glad to send his survey fossils to Yale for identification.

Marsh, King, and Powell frequented the same social milieu in the mid- to late 1870s—the Century Club in New York and, later, the Cosmos Club in Washington. They thus had plenty of opportunities to confer and develop political strategies. There's little documentary evidence of such conferences, although Hayden's 1874 complaints about "certain parties" probably referred to them. Neither Marsh nor Powell was prone to written confidences, and King reserved his for fellow patricians like Adams and Hay. Still, events that would occur rapidly and shockingly for the Hayden faction at decade's end show that the three men had been busy.

8

Cope the Explorer

Cope was understandably hesitant about another trip west after his 1872 difficulties. "Last week I did but little as I had a threatening of a touch of my Bridger fever," he wrote to his father in January 1873, and complained that he was so weighed down with bread-and-butter writing jobs that he had trouble "getting at my Wyoming fossils." He applied to the pending Northern Pacific Railroad survey, which would have taken him outside Marsh's widening influence, but it fell through because of the Indian situation. "I do not know if I shall go West this summer," he wrote in May. "Hayden objects to the expense of sending me this year, but he is a wire-puller and I suspect has other reasons."

In July, Cope did go, but he cautiously stayed with Hayden's survey and worked in known areas. "Hayden only gave me $250 as he is very short of funds this year," he told his father, apologizing for his sponsor's unreliability. He tried to keep out of Marsh's way, although on July 18 he reported that he'd seen the Professor "running about in some excitement" at the Cheyenne railroad station. It's easy to imagine Edward sinking involuntarily into his seat as Othniel bestrode the sunny platform. Although Marsh later became tubby, he was a tall, impressive figure in 1873, with a hard, mocking gaze. "He is said to be gone west of Bridger," the letter added placatingly. The reaction of the straight-laced Alfred Cope to his son's *American Naturalist* brawls can be imagined.

Cope started from Greeley, Colorado, with the same kind of shoestring outfit he'd had the year before, although the riffraff didn't sink to Fort Bridger depravities. "I have had passable success with the men and animals," he told Alfred, "but the disagreeable part of this business is the necessity of associating with such men as one has to employ. It is almost enough to prevent me from undertaking it." Still an avowedly devout Friend, he read aloud from the Bible at night and forbade swearing in camp, which must have seemed "hi toned" enough to the riffraff. They horsed around or did camp chores while Cope recited. The pious little party headed east to the Platte River headwaters, where they dug up quantities of creatures like those Leidy and Marsh had been describing for years, although these were different enough for Cope to call them "quite distinct" from Nebraska's or Wyoming's.

Although it was not a spectacular trip, it was the beginning of Cope's major fossil-hunting expeditions. The younger man's vision of natural diversity kept driving him west despite both adventures and misadventures that sometimes resembled Buster Keaton's more than William S. Hart's. Cope was more of an explorer than Marsh. Othniel went after things he knew were there: material for anecdotes and fossil deposits likely to yield useful material. Once the lively aspects began to pall, he saw little reason to endure months of labor and discomfort when he could pay others to get what he wanted. Cope sought the unknown, a more risky goal, but a compelling one, since few thrills rival that of reaching it. He would find a number of unknown evolutionary worlds, although they would prove harder to recognize than birds with teeth—harder for Cope as well as for other paleontologists, not to mention the public.

Cope just couldn't resist those tantalizingly little-known Eocene formations near Fort Bridger, so he made a furtive trip there in October to pick up more *Loxolophodon* bones. He was beginning to think there might be something remarkable in the Rockies farther south, maybe a closer link than was yet known between the Tertiary

Period, the "age of mammals," and the Cretaceous, the end of the dinosaur age. The apparent discontinuity between the two was one of paleontology's major problems, because it seemed to indicate that there was no connection between the dinosaur age and the age of mammals. It was as if life had changed through Cuvierian creation rather than through Darwinian evolution. Cope had held the Cuvierian view in an 1867 article, in which he wrote that "a great change of temperature" at the end of the Cretaceous had destroyed all animal life. "Then began again . . . the introduction of entirely new forms of animal life more like those of modern times," he concluded, calling the early Tertiary " 'the morning of the sixth day' in the Mosaic record of the Creation."

But his trips to the West had changed Edward's mind by 1873. "At this place there is a great puzzle as to the limits of the Cretaceous and Tertiary formations," he wrote to his father from Fort Bridger on October 3. "Indeed, this whole country seems to furnish closer connections between the two than any other . . . Last summer I made some important progress in proving that what many called Tertiary was really Cretaceous." An attack of Marsh fever distracted him from these reflections—"Marsh was wroth because I didn't credit him with the discovery . . . So if you do a good thing it is either not true or done long before! etc."—but he returned to them the next year. "I sum up in the new *Bulletin* the results of the relations of the Cretaceous and Tertiary formations in their points of contact," he told his father the next January. "The result is that the beds gradually pass from one to the other, but there is an interruption in the history of life. But for animals and plants the period of interruption *is not the same;* the Tertiary plants were introduced long before the Tertiary animals! So there could have been no such great destruction after all, and it goes to prove that change took place in both cases by migration and not re-creation."

Cope saw, in other words, that the massive change from the giant saurian world of the Cretaceous to the giant mammal world of the Tertiary was only a partial one, because the same vegetation had characterized both: the same forests of oak, fig, palm, and other existing trees. Such forests had appeared in the Cretaceous, when dinosaurs still ruled. If plants linked the two periods, then animals perhaps did also. There may have been a time when mammals even more primitive than Leidy's Eocene lemurs and uintatheres had lived, and Cope suspected he would find their fossils in unexplored badlands far to the southwest of Fort Bridger. It would be important evidence—more far-reaching than toothed birds—that geological time proceeded not in Cuvier's way but in Darwin's. It also would be beyond Marsh's reach.

Yet it was one thing to dream of unknown worlds across the Rockies and another to find them. Marsh's minions continued to pose obstructions. In December of 1873, one of them wrote to the Professor that Cope had invaded his territory again, and had left "*marks* distinctly made, I presume . . . so I judged he could strike certain places next spring. These marks are now defaced by the sole of my boot, but I know where they are." There was also the question of financing, which Ferdinand Hayden seemed unlikely to provide. "That he has promised often enough but, as in many other cases, not performed," Cope had told his father in December 1873. "I have to furnish him with backbone and keep enough on hand for home supply. He is terribly beset by his rivals in Washington, some of whom have no love for me, as I have flanked them often on publications. So they complain of me to him, and he, not knowing the facts, comes down on me."

Fortunately, Hayden, King, and Powell weren't the only surveyors. The Army Corps of Engineers was running a survey beyond the 100th meridian, and its commander, Lieutenant George Wheeler, was not anti-Cope. "Wheeler's and Hayden's surveys are opponents again this winter," Edward wrote to Alfred in March 1874. "Whichever succeeds, I will try to get a salary or leave them!" In July, he succeeded in drawing up a contract with Wheeler "at the rate of $2500 per an. and $30 per mo. additional for provisions when out in the field, and all expenses of expedition paid." It was "very different from any arrangements hitherto made," a step up from Hayden's penny-pinching and bumbling. "It is pretty clear I did well in accompanying Wheeler's survey," he wrote to Alfred from Denver, "since the points Hayden had in view are too near the hostile Indians to be safe. At present I go away from these difficulties to the west."

Less fortunately, Wheeler differed from the indulgent Hayden in expecting Cope to work as expedition geologist, not just as a bonehunter. "I am only scientific, not personal or financial director hence have only partial management . . . ," he complained from Taos, New Mexico, "and so I have to delay at places most unprofitable to me against my will . . . All this comes from the system of orders and regulations which it is customary to issue for the government of parties of engineers but which is useless for explorers for unknown objects in new fields. It is absurd to order stops where there are no fossils, and marches where fossils abound! We are breaking through this, however, and will lay it aside before long."

It's unclear who was Cope's *we,* but he lost little time in "breaking through" regulations. In mid-August he learned from a local collector, a Catholic priest, that his hopes for very early mammal fossils in the San Juan Basin west of Santa Fe were well founded. The priest's fossils included *Bathmodon,* a mammal that Cope had found at Fort Bridger and had likened to a combined elephant, tapir, and bear. "I could not have desired anything more agreeable and not surprising than this discovery of our lowest Eocene 500 miles south of where known!" he wrote to his wife, who was waiting in Denver, no doubt with nursing supplies. "And in full confirmation of my theory that the fauna of Fort Bridger beds comes from the south!"

Less high-mindedly, Cope added that Annie "need not mention this for awhile," and explained how he planned to circumvent the authority of his superior, a zoologist named H. C. Yarrow, by going over Lieutenant Wheeler's head. His next letter was exultant: "I have just returned from consultation with Gen. Gregg at his headquarters," he wrote three days later. "Dr. Yarrow presented his case and his written instructions, and I then explained to him the impossibility of fulfilling the instructions and succeeding in my paleontological work. To my delight the Gen. at once took my view of the case and set the Dr. at liberty to violate and disregard the points which I have found so objectionable . . . Every thing will I hope go on swimmingly."

Things didn't go on swimmingly at first. Dr. Yarrow, clinging to shreds of authority, objected when Cope proposed to divert the party to explore the San Juan River headwaters "for the purpose of discovering the Eocene lake formation that I knew was there after I saw the teeth weeks ago at El Rito." All members of the party spent several weeks pottering around more recent formations, "moderately rich in fossils" of dogs, horses, camels, mastodons, and rhinos, and Cope reined in his impatience by being interested in everything, as usual. He listed wildflower species and observed the local Pueblo Indians, whom he found friendly and intelligent. He wondered whether their civilization had been imposed by the Jesuits or was indigenous. "I am not sure but evidence favors the former view," he told his father, matching Marsh's obliviousness of "the origins of the Indian race."

In early September, Cope finally overrode the hapless Yarrow's objections and went to look for his Eocene goal, taking two guides, a collector named Shedd, and a pack horse across the Gallinas Mountains to "where we crossed a low ridge and looked westward far into the flat basin of the San Juan Tributary of the great Colorado." There he saw formations that "were nothing less than the long looked-for bad lands of the Eocene! My dream of several years was realized." After reaching the badlands on the next day, he wrote, "we began to find fossil bones! The first thing was a turtle and then *Bathmodon* (Cope) teeth! and then everything else rare and strange until by sundown I had 20 spec. of vertebrates! all of the lowest Eocene, lower than the lowest at Fort Bridger. The most important find in geology I ever made, and the palaeontology promises grandly."

A few days later, however, Cope received an order from Wheeler to return to the main party. "I feared I was to lose the result of my hard work," he wrote to Annie, but he obediently returned after caching the fossils he'd already collected. To his surprise, Yarrow had left for Washington, and Wheeler gave him men, mules, and provisions to continue his fossil-hunting. A member of the party had accidentally killed himself with his own revolver, and Wheeler had sent Yarrow back to deal with that. Edward at last had full control. "I have now a good outfit on my hands and will make the most of it," he told his father on September 14. Proudly affirming his prediction of fauna even older than Fort Bridger's, he told Alfred of finding forty-three vertebrate species in five days. "The situation thus fulfills my expectations exactly," he exulted, "and nothing remains but to work it up carefully."

A week after he returned to the badlands, Edward told Annie that he had over seventy-five vertebrate fossil species, mostly new. "The weather is lovely," he continued. "Days warm and nights cold. I usually have ice water to wash in in the morning, and after breakfast a ten-mile ride to the bluffs to work. I have a grand appetite and am getting fat, fatter than I have been since the fever days at Fort Bridger."

Cope was right about the importance of the San Juan badlands. Because of his notes, they are among the most fully documented of early paleontological sites. The New Mexico Museum's Spencer Lucas was able to show me exactly where Cope's party had descended through the pinyon-juniper woodland of the Gallinas Valley and seen the maroon-, brown-, and yellow-banded Eocene badlands to the south. "Cope knew by the succession from Triassic to Jurassic to Cretaceous strata he was crossing in the Gallinas Valley that he was getting near to the Eocene," Lucas said. "When he saw the badlands he would have recognized them from formations he'd already seen in southwest Wyoming."

As we walked around the bluffs, Lucas pointed out bits of turtle shell and bone from Cope's *Bathmodon* (today known by a prior name, *Coryphodon*). "This is how Cope would have worked here, walking around the base and looking for material that had washed out, then following them uphill. It's easier to go up these things than get back down." I believed him, looking up at the steep jumble of mudstone and cave-like erosion holes decorated with mud stalactites. Then Lucas indicated a scatter of objects that looked like onyx; they were gar scales that had fallen from a cylindrical fossil fish embedded below one hole. "That might be worth collecting," he said. "What we have here is a cemetery, a necropolis for vanished creatures, and we're like priests who try to reincarnate them." He gestured at a pyramidal bluff. "That represents about a million years of deposition in the early Eocene of fifty-five to fifty-seven million years ago. Bertrand Russell once said history is the only reality. Cope found a new piece of reality here, knew what he'd found, and was absolutely correct about its significance."

Cope discovered even more in the San Juan Basin. Under the reddish stone of the early Eocene, he came upon a grayish-green mudstone, which he named the Puerco Formation, after the local river. He thus discovered about eight million years of geological time, since that formation was the first evidence in North America of an entire epoch between the Cretaceous and Eocene, 65 to 57 million years ago. "This Paleocene horizon," Osborn wrote, "contained a truly archaic mammalian fauna, scarcely dreamt of . . . representing with incomparable richness and profusion the formerly luxuriant life of this now terribly arid land. With the exception of one small herbivore . . . every tooth and every bone of this bizarre fauna was new to science." According to Björn Kurtén, a modern paleontologist, it provided "the first glimpse of vertebrate life after the extinction of the dinosaurs."

Spencer Lucas showed me the Puerco beds when we were on our way back to Albuquerque. As we neared the town of Cuba, the colorful Eocene badlands gave way to a stratum of sandstone, then to the soft gray mudstone Paleocene formation. "Cope probably never heard the word Paleocene in his life, though," Lucas said. "A Swiss paleobotanist named Wilhelm Schimper coined it in 1874 to refer to a new plant fauna in what was then thought of as the early Eocene, but the name didn't come into general currency until the 1920s."

Unfortunately, the Puerco beds that Edward explored in 1874 had no animal fossils, just petrified wood. "I have been at work on the bad lands," he wrote to Annie, "and very hard work it is; much harder than last year. We go for a mile or more and see not a fossil." Not until 1881 would Cope learn what had lived in his new epoch. That was the year he hired a sourdough collector named David Baldwin to work the San Juan Basin's western side. Marsh had employed Baldwin to explore the basin after Cope's 1874 discoveries, but had let him go, because he thought his fossils uninteresting. Equipped with little more than a mule and a pickaxe, Baldwin proceeded to collect over a hundred species for Cope, "nearly all of them new to science."

Cope ran into a basic problem with his explorations after the unknown. It took him years to figure out what he'd discovered. There was no immediate recognition, such as Marsh had enjoyed with his toothed birds. Cope had to invent new biological categories just to classify his fossils. For example, he found many predators "whose size and powers of destruction equaled those of the bears, lions, and tigers of modern times," but whose bones were more like a shrew's; he therefore made up a whole new order called the Creodonta—"flesh-toothed"—for them. Taxonomists since have changed many of Cope's early mammal classifications, but they were fruitful ideas. Less famous than dinosaurs, the extinct mammals have more evolutionary significance to the living world, because they were the forerunners, if not the direct ancestors, of today's mammals. I saw tray after tray of their bones at the New Mexico Museum, from rhino-size *Coryphodons* to dog-size creodonts to early primates so small that their teeth were barely visible. "There's so much here," Spencer Lucas said, "that I wonder why Cope didn't come back to do more work. He could have spent the rest of his career here."

Edward was a contented paleontologist in late 1874. "We have done a great deal since I last wrote," he told Annie on October 11. "The present camp is . . . in a small rolling plain with a circle of bad lands all around us . . . We have Navajo Indians passing by us as we are near the trail from Abiquin to Canon Largo. The Utes also call on us now and then, and two families have taken a fancy to camp close to us. We expect to have plenty of fresh meat now, as we have plenty of coffee, tobacco, and flour to spare for barter. We have plenty of wild fruits and nuts to vary our larder . . . Tonight we sat around a fire of pinon wood. I am now writing by its light stretched out on the ground."

But he was never contented for long. "I have discovered the deposits of another fresh water lake of much greater age, say Cretaceous, not many miles from here," he continued, "which contains remains of Saurians, one like *Laelaps* I have a tooth and vertebra of . . . If I have time I will endeavor to find some good place where remains are abundant and perhaps find a whole population of Saurians such as occur in Kansas, with the difference that these are of the land, the latter of the salt sea." Land saurians—dinosaurs—still were largely unknown in 1874. Paleontologists had yet to discover dinosaur beds as rich as the mammal ones. To Cope, finding the "Saurians" near the unprecedentedly old Tertiary formation of the Puerco must have been like digging down to the basement of a ruined palace and breaking through the floor to a huge dark chamber underneath.

Two years elapsed before Edward could seek his "whole population of Saurians," however. He told Annie that Wheeler was "greatly pleased with my results," but the army never hired him again. The punctilious Wheeler also barred him from using his survey's publications to attack Marsh, although Marsh complained about Cope's reports anyway. And Cope didn't get west at all in the summer of 1875; he was busy writing up his "enormous collections of the previous four summers." That fall, his father died, which must have been both sad and liberating. Edward lost his main intellectual confidant and stabilizing influence, but he gained an inheritance of about $250,000, which would grant him a degree of freedom from scientific bureaucrats. He moved out of Haddonfield and bought two adjacent houses on Pine Street in Philadelphia, one for his family and one for his fossils. Alfred's death also freed him from the overwrought piety that seems to have been his way of dealing with filial guilt. He soon left the Friends and adopted Unitarianism.

Now Cope could compete with Marsh in hiring field collectors. He tried again to lure the former professor Mudge away from his rival, forcing Marsh to pay Mudge even when he wasn't working, a situation that must have chagrined the thrifty Othniel. Then Cope found Charles Sternberg, the man who later would recollect Edward's paleontogical visions. Sternberg had developed a similar visionary outlook, mixing frontier religiosity with a basic geological education, and had decided to make fossil-hunting his vocation. His autobiography told how he had dreamed of a deposit of five-million-year-old fossil leaves on the Kansas prairie and then had gone right to the spot—and found them. "Go back with me, dear reader," Sternberg wrote, "and see the treeless plains of today covered with forests . . . we can imagine that the Creator walked among the trees in the cool of the evening . . . but the glorious picture is only for him who gathers the remains of those forests, and by the power of his imagination puts life in them."

Since Mudge had rejected the inexperienced Sternberg's application to work for Marsh, the young man applied to Cope in 1876. Edward promptly sent him $300 for supplies, which, Sternberg wrote, "bound me to Cope for four long years and enabled me to endure immeasurable hardships and privations in the barren fossil fields." The hardships began promptly that spring, as he competed with Marsh's parties in western Kansas. Afraid the Marsh riffraff would jump his claim if he went away to eat, he once lived on parched corn for three days while cutting eight hundred pounds of mosasaur bones from the chalk with a butcher knife.

Sternberg probably was relieved to be called away in August to help Cope and another assistant, J. C. Isaac, look for the mysterious "land Saurians." His relief must have faded when he learned where they were going. Instead of returning to the San Juan Basin, Cope planned to search in the Judith River badlands of north central Montana, the area from which Hayden had sent Leidy dinosaur fragments in the 1850s. It was a startling destination in 1876. Sitting Bull, Crazy Horse, and their two thousand associates had just extirpated Colonel Custer less than two hundred miles to the southeast, and white panic pervaded the plains. J. C. Isaac had watched Indians kill five of his companions in Wyoming earlier that year, and in June Sternberg had seen intimidatingly wide trails of Cheyennes moving through Kansas to join Sitting Bull. "All was excitement, and the professor was strongly advised against the folly of going into the neutral ground between the Sioux and their hereditary enemies, the Crows," he wrote. "A member of either tribe might kill us, and lay our death to the other tribe."

The usually cautious Cope played down the Indian danger, especially in letters to his wife. "There is no risk from the Sioux," he wrote from Helena on August 14, six weeks after Custer's death. From Fort Benton five days later he added, "Cock-and-bull stories about the Sioux have been published in the Helena and other papers . . . Every one considers them false." Cope thought they'd be safe, Sternberg recalled, "since every able-bodied Sioux would be with the braves under Sitting Bull, while their squaws and children would be hidden away in some fastness of the mountains. There would be no danger to us, he reasoned, until the Sioux were driven north by the soldiers who were gathering under Terry and Crook for the final struggle. Judging from past experience, he concluded that we would have nearly three months in which to make our collection in peace." Cope's men, though, had cause to suspect his judgment. "When we first met him in Omaha," Sternberg wrote, "he was so weak that he reeled from side to side when he walked." Cope's headlong perseverance might have brought to mind Custer's dash to his fate.

Western hardships again did Cope good, however. "I begin to feel like a camper already," he wrote to Annie after a four-day stagecoach ride to Helena with a driver who sometimes was so drunk, he could hardly stand. "For the first time in six months I had an appetite for dinner. Exploring will doubtless have its usual effect on my health." From Helena they took another coach, which "nearly upset several times." The driver got lost in the dark, and when he found the road, they "had a tremendous climb over a mountain where we nearly all had to walk." That was just to reach Fort Benton, where Cope hired the outfit they needed. It took them another week to get to the badlands.

"Since my last I have got to work in regular style," he wrote on August 27. "I am close to my field with regular outfit and everything working nicely . . . We have just been reading the Bible and prayer . . . Sternberg is a very religious character and Isaac is a good fellow and I am fortunate in having secured a very respectable man for a cook. They behave differently from my Colorado party who scrubbed their teeth and chopped wood while I read." The scenery was "very fine," with "splendid grass as far as the eye could reach . . . It abounds in buffalo, antelope, deer, wolves, Indians etc. . . . and now in sight of our camp on the Missouri is a large camp, say some 30 lodges of some Indians I do not know."

Cope didn't send this letter right away, fortunately for his wife's peace of mind, but continued after he'd visited the unknown Indians the next day. They turned out to be a thousand peaceful Crows. "Last night," he added, "a chief of the River Sioux, Beaver Head, and his squaw, slept in our camp under the wagon, and took breakfast . . . They behaved with perfect good manners, using their knife, fork, and spoon for everything. I showed them your photo, at which Beaver Head said 'pretty good' and the squaw suggested that thee cried when I went away. In the large trading post opposite to us, the men say there is positively no danger from any Indians." In his next letter, a week later, Edward wrote that he'd had four Crow chiefs to dinner. "They were greatly amused to see me take out my teeth and put them back," he added proudly. "One man (Assinibonie Jack) rode several miles to see it."

They discovered plenty of land saurians. "The bed that I came for I have found," he wrote on September 10. "It is supplied in many places with many shells and teeth, but for considerable intervals one finds nothing. There are also many huge bones, but they are generally alone, and badly broken so that they are not worth digging out. I find so far nothing but reptiles and fishes, up to this time 30 species of which half are new." The fossils' fragility forced the men to improvise by covering bones with cloth soaked in boiled rice paste to stop breakage. This remains standard practice, with plaster instead of rice paste, and Cope has been credited with its invention, although Marsh told Charles Schuchert in 1892 that *he* had invented it from watching doctors set broken bones in casts.

None other than Schuchert wrote that Cope's 1876 trip for the first time "made it clear that dinosaurs had been present in the West in considerable abundance in Cretaceous times, although his material gave little hint of their skeletal structure or of their immense size." The vision that had led Edward to the San Juan Paleocene worked again, but, once more, novelty prevented quick recognition. The fossils puzzled him even more than the San Juan Basin ones had. "We find in the high rocks here many bones and teeth of huge fossil reptiles like *Laelaps* and *Hadrosaurus*," he wrote to his daughter, Julia, in September. (Although she was only twelve, she was replacing Alfred Cope as her father's intellectual confidante, a role she would play for the rest of his life.) "They are as large as elephants, and their teeth are very small, no larger than my little finger . . . Then there are swimming lizards' and turtles' bones and a kind of animal with the teeth packed together like the pavement on the back hall of Fairfield, and I don't know whether it is a fish or a reptile."

The pavement-toothed fossil turned out to be a ray, a common fish, but many of the other finds were strange indeed. The strangest were the ceratopsians, the rhino-like, horned monsters that now, in countless illustrations, battle tyrannosaurs but that were unknown in America before that summer. Cope had found ceratopsian rib and leg bones in Wyoming in 1872, but they were not enough to give him an idea of what the dinosaur had looked like, although he'd named it anyway—*Agathaumas sylvestris*—"marvelous forest creature." Now, he learned from the Judith River fossils that the creatures' heads had been decorated with long sharp horns, so he named this genus *"Monoclonius,"* because it had a single large horn on its nose. "I assisted him in digging out his specimen of *M. cras-*

sus, a species distinguished by a small horn over each orbit, and a large one on the nasal bones," Sternberg wrote, "and I myself discovered two species new to science. One of these, *M. Sphenocerus,* was six or seven feet high at the hips, and, according to Cope, must have been twenty-five-feet long, including the tail."

"On the whole, we have occasion to be satisfied with our situation," Cope told Annie. "The cook's name is Austin Merill, and he is the best I ever had . . . I hired a guide familiar with the country who keeps us in fresh meats. His name is James Deer and he is a clever fellow. Altogether, my camp is the best I ever had."

Yet shadows lurked behind the sunny picture. Some of the rental horses were so unruly that they had to be clubbed into submission with a whip handle. The badlands water was muddy and alkaline, and Cope complained that the cook's fancy dishes caused indigestion, "to the detriment of our dreams when we come home at night." That was putting it mildly. "Every animal of which we had found trace during the day played with [Cope] at night, tossing him into the air, kicking him, trampling upon him," Sternberg recalled. "When I waked him, he would thank me cordially and lie down to another attack. But the next morning he would lead the party and be the last to give up at night. I have never known a more wonderful example of the will's power over the body."

The Judith River bluffs are much higher than the White River's, and the canyons are so steep and convoluted that disorientation and injury were constant dangers. J. C. Isaac vanished one evening after becoming "bewildered among the hummocks and ridges" and reappeared only as Cope and Sternberg were setting out to look for him the next day. Another time they all narrowly escaped falling into a chasm in the dark. "At night the view from above of these intricate passages was appalling," Sternberg wrote. "The black material of which the rocks are composed did not permit a single ray of light to penetrate the depths below, and the ebony-like darkness seemed dense enough to cut."

Ominously, the friendly Crows had packed up and left soon after the expedition arrived, and even more unsettling departures ensued. When Cope returned to camp one day in mid-September, his cook was gone. "Cause, fear of Indians," he wrote to Annie on October 3. "Then my worthy guide Jas. Deer announced his intentions of doing the same; cause, chance to make more money—but I don't believe that to be the real reason . . . So we are reduced to three." Cope told Sternberg they had deserted even though he'd already paid them for three months. "It seems," Sternberg recalled, "that the scout had come across Sitting Bull's war camp, where thousands of warriors, drunk with the blood of Custer and the brave men of the Seventh U.S. Cavalry, were defying the Government . . . The camp was only a day's journey from us, and the scout and our valiant cook had concluded that their precious scalps were too valuable to risk."

Despite his men's fears, Cope stayed for almost two more weeks, until the river steamer was due to make its last departure of the season from the trading post at a place called Cow Island. "Yesterday, we moved out of the bad lands after procuring some good things," Edward continued in his October 3 letter. "We had a difficult task to get down to the Missouri. Had to let the wagon down by ropes." During one such maneuver, wagon and team together tipped over and crashed downhill, revolving completely three times before coming to rest upright on a ledge "as if nothing had happened."

They had accumulated seventeen hundred pounds of fossils, but that wasn't enough for Edward, who led Sternberg back into the badlands for more. While they were returning toward the river camp late in the evening, they took a trail that ended in a six-hundred-foot cliff, and discovered that there was a maze of ravines between them and the camp. "I knew the uselessness of trying to combat his iron will," the harried Sternberg remembered, "but I pleaded with him against the folly of attempting to thread in the darkness those black and treacherous defiles, where a single misstep meant certain death." They threaded the defiles anyway, and finally reached the water six miles upstream from their camp, but "the river at several points struck the bluffs so that we could not follow it, but had to climb the precipice again." After hours of scrambling up and down bluffs in the dark, they found themselves opposite Cow Island, so they crossed the river the next morning and had breakfast with the soldiers there.

The steamer had arrived, but the captain wouldn't stop at their camp four miles downstream for fear of getting stuck. Cope bought a "Mackinaw" flatboat that was lying around, and they rowed it downstream the next morning with the intention of floating the fossils back the next day, when the steamer was to leave. But when they got back to camp, rowing the heavily loaded boat back upstream proved impossible. It was Cope's iron will that saved the day. "We took our strongest horse, Major," he wrote, "and hitched him by our lariat ropes 180 feet long to the boat, and so hauled it along *à la* canal boat. After sundry adventures in which we all got very wet, and the horses rolled down the bank into a mud hole, we reached Cow Island in good case, having been three hours in coming up the river." Sternberg described Cope as plastered with mud, his clothes in rags, but the next morning he steamed away nattily in his last intact garments—a summer suit and linen duster. "So all ended well," he wrote to Annie, "and I am now going down the Missouri to Yankton to meet my missing self, who is much wanted by her husband."

Edward's scanty wardrobe didn't undermine his sense of propriety. "We had good weather, made good speed, and the officers of the *Josephine* are all nice and intelligent men," he wrote from the steamer on October 21. "There was no liquor on board, and the amount of swearing was small." Activity was feverish on the banks as army units, including newly recruited replacement companies of "Custer's unfortunate regiment," prepared to chase Sitting Bull. "The debarkation and marching from the bank opposite Fort Lincoln was a pretty sight," he wrote, "and the officers' wives watched from our steamer their departure. None knew when they would ever see their husbands again, but they were cheerful, some too much so, but some showed their feeling."

Despite such sobering reflections and his seventeen-hundred-pound haul, Cope still didn't have enough bones. Perhaps he'd envied Marsh's grave-robbing, because when the boat passed a mortuary scaffold, he "gathered a number of skulls and skeletons of Sioux with a bag of tools buried with a chief and brought them on board and boxed them up." He didn't have the autocracy over his fellow travelers that Marsh enjoyed, however, and ordinary frontiersmen had finer sensibilities about dead Indians, at least, than did Eastern scientists. "The uproar it created among the poor white element that run the lower offices," he went on, "scared the Captain so that he ordered them all taken back to the place where I'd obtained them . . . The engineer declared that the Sioux would capture the boat! I was wroth . . . It is not a nice job, taking dirty skulls from skeletons not carefully prepared, and it is done at some risk to life."

Edward was at less "risk to life" than his two assistants, who'd stayed in the badlands to dig up more bones and then return the outfit to Fort Benton. One horse amused itself by running at a clifftop with Sternberg on its back and trying to pitch him to death by stopping at the brink. When Sternberg and Isaac finally got back to Fort Benton, in front of an early November blizzard, they heard that all had not "ended well" for the soldiers at Cow Island. Sitting Bull had crossed the Missouri there on his way to Canada and had wiped them out. Sternberg seems to have shared the Western tendency to demonize Sitting Bull as an omnipresent avatar of the cruel savage, however. According to Robert M. Utley's 1993 biography of the Hunkpapa leader, he and his much reduced band had been hundreds of miles southeast of the Judith River badlands when Cope's cook fled in late September, and in November they were still about a hundred miles to the southeast. Of course, Sternberg couldn't have known all that as he toiled across the autumn plains.

In the fall of 1876, Cope may have felt that he lived a charmed life. He had had unprecedented summers of fieldwork, finding not only what he'd hoped for, but what he'd not even imagined. He had not succumbed to Indians or mountain fever, and, best of all, had not run up against Marsh's pitchfork. When he invited Sternberg east for the winter to work with him on preparing the fossils, Charles described a happy man. "I had a standing invitation to eat dinner every Sunday with the Professor and his wife and daughter, a lovely child of twelve summers," he wrote. "The Professor's conversation was a feast in itself. He had a wonderful power of putting professional matters from his mind when he left his study, and coming out ready to enter into any kind of merrymaking. He used to sit with sparkling eyes, telling story after story, while we laughed at his sallies until we could laugh no more."

Yet Marsh's pitchfork had not vanished, as Edward well knew from following the Professor's Red Cloud campaign in the newspapers. On July 23, 1875, he wrote to his father, "Dr. Hayden states that parties whose characters have been aspersed intend to push the matter to a full vindication of themselves at any cost . . . The *New York Independent* has an editorial against O. C. Marsh in which it says that some of his charges are undoubtedly false." Edward would be disappointed if he thought politics would slow Marsh down. With the new Peabody Museum to house his collection and his brilliant assistant, Oscar Harger, to help interpret the mass of fossils, the Yale Professor was just getting started.

9

Huxley Anoints Marsh

Marsh's small army of collectors had swarmed across the West again in 1875, this time with a list of instructions for digging, packing, and posting their finds. "Get all the bones of every good specimen if it takes a week to dig them out," was the seventh of Othniel's fifteen commandments. "The loss of a single tooth or toe bone may greatly lessen the value of a specimen." Marsh was so eager to get good collectors that he invited Samuel Williston, a young college graduate from Kansas and a naturalist, to come to New Haven for a month's training before the 1876 season. Williston had to pawn his watch to pay for his fare.

"My heart was in my mouth as I knocked on the basement door of the old Treasury Building and heard a not very pleasant invitation to come in," Williston recalled. "There was a frown on Marsh's face, accentuated by his nearsightedness, as he asked for me to state my business . . . But he quickly made me feel more at ease. He found me quarters in a little building at the rear of the Peabody Museum, then approaching completion. The next day he set me to work studying bird skeletons with Owen's comparative anatomy as a guide."

Williston, inspired, returned to Kansas and a month later described a fossil bird search to Marsh: "I followed your directions strictly—marking out the spot and looking over it inch by inch lying flat on the ground, and afterwards working over the wash and loose soil as thoroughly with the knife. I think we have gotten *every* fragment . . . I looked for them as if they were diamonds." Marsh also imbued Williston with zeal for keeping Cope's collectors at bay, one reason that Charles Sternberg had such a hard time that summer. Williston had assimilated his employer's hard-boiled approach, and the mystical Sternberg was no match for him. Williston, who would return to Yale the next winter as one of Marsh's laboratory assistants, eventually became a professor of paleontology at the University of Chicago. Sternberg would remain a hardscrabble bonehunter.

Marsh's collectors eventually provided him with parts of fifty specimens of the toothed Cretaceous bird *Hesperornis* that he had discovered. This avian arsenal allowed him to make the best demonstration of an evolutionary transition since the 1862 discovery of *Archaeopteryx,* and Marsh's toothed birds were a more convincing link between reptile and bird in 1876, because *Archaeopteryx*'s teeth would not be discovered until 1877. (The "*Archaeopteryx* head" Cope had mentioned in 1863 evidently belonged to something else.) When Marsh's monograph on the toothed bird was published, T. H. Huxley wrote that his discovery "completed the series of transitional forms between birds and reptiles, and removed Mr. Darwin's proposition . . . from the region of hypothesis to that of demonstrable fact." Even the anatomist Richard Owen, Huxley's enemy, called the monograph "the best contribution to natural history since Cuvier."

Of course, Marsh had more than toothed birds in his arsenal. His collectors had supplied him with enough prehistoric horse bones to show in detail how the equine line had evolved over some fifty-five million years, from cat-sized browsers in Eocene forests to big grazers on Pleistocene prairies. No paleontologist of the time—indeed, no naturalist of any kind—did more than Marsh to validate Darwin's "transmutation," and influential evolutionists like Huxley and Harvard's Asa Gray appreciated it. In 1874, Gray offered Marsh a Harvard professorship, which he gracefully declined.

It's unclear exactly how Marsh developed the modest ambition of substantiating evolution with his fossils, but he told a rousing, patriotic anecdote about its origin. "When a student in Germany some twelve years ago," he wrote in 1877, "I heard a world-renowned professor of zoology gravely inform his pupils that the horse was a gift from the Old World to the New, and was entirely unknown in America until introduced by the Spaniards. After the lecture, I asked him whether no earlier remains of horses had been found on this continent, and was told in reply that the reports to that effect were too unsatisfactory to be presented as facts of science. This remark led me, on my return, to examine the subject myself, and I have since unearthed, with my own hands, not less than thirty distinct species of the horse tribe in the Tertiary deposits of the west alone, and it is now, I think, generally admitted that America is, after all, the true home of the horse."

The Professor's assistants would have smiled at the part about his unearthing thirty horse species with his own hands, but the anecdote's timing has a ring of truth, given Marsh's penchant for long-term strategies. There's no doubt that he amassed the world's largest collection of horse fossils in the decade after his first trip West, enough to awe T. H. Huxley when he visited Yale in August 1876. Huxley was the man to awe, since he was largely responsible for turning biology into a profession. Starting as a lowly surgeon's assistant in the 1846–1850 *Rattlesnake* expedition to Australia, he had transformed himself in the next decade from an amateur authority on jellyfish to a professor of natural history at London's prestigious Government School of Mines, and thus had been in a pivotal position to support Darwin after the publication of *The Origin of Species* in 1859. By 1876, he held several other academic and administrative posts, and his position in the new evolutionist establishment was not unlike that of the Archbishop of Canterbury in the Church of England. "Huxley's Church Scientific, with its canonizations and demonizations, invited a sublimated form of worship," wrote Adrian Desmond, his biographer. "Huxley relished his role as Darwin's protector."

Huxley considered himself a horse authority at the time. A young Russian paleontologist, Vladimir Kowalevsky, had traced a sequence of Old World horse ancestors back to the Eocene, and Huxley, like most evolutionists, believed they had evolved mainly in Eurasia. His supposedly "state of the art" horse sequence proved skimpy in comparison with Marsh's, however. The Old World horses consisted of a few genera linked by limited characteristics. Marsh trotted out genus after New World genus, showing the gradual transmutation from four-toed creatures of Eocene and Oligocene forests to three- and two-toed Miocene and Pliocene genera to single-hoofed horses of the Ice Age grasslands.

"At each inquiry, whether he had a specimen to illustrate such and such a point or exemplify a transition from earlier and less specialized forms to later and more specialized ones," wrote Huxley's son Leonard, "Professor Marsh would simply turn to his assistant and bid him fetch box number so and so, until Huxley turned upon him and said, 'I believe you are a magician; whatever I want, you just conjure up.'" Far from feeling put out by this display of New World priority, the formidable Huxley was charmed. "My excellent host met me at the station, and it seems he could not make enough of me," he wrote to his wife. "I am installed in apartments which were occupied by his uncle, the millionaire Peabody, and am as quiet as if I were in my own house . . . He is a wonderfully good fellow, full of fun and stories about his western adventures, and the collection of fossils is the most wonderful thing I ever saw."

Marsh described the encounter with disarming modesty. "My own explorations had led me to conclusions quite different from his . . . ," he wrote. "With some hesitation, I laid the whole matter frankly before Huxley, and he spent nearly two days going over my specimens with me, and testing each point I made. He then informed me that all this was new to him, and that my facts demonstrated the evolution of the horse beyond question, and for the first time indicated the direct line of descent of an existing animal. With the generosity of true greatness, he gave up his own opinions in the face of new truth, and took my conclusions as the basis of his famous New York lecture on the horse."

Huxley's three lectures on evolution in New York's Chickering Hall in September were front-page news, drawing as much press as a yellow fever outbreak in Georgia. Most papers reported on them favorably, including the *Herald,* although it vacillated uneasily in its attitude toward the Darwinist prelate. "There are few English writers in any department of letters who possess such rich gifts of expression," an editorial fawned, but then complained that Huxley had "seen fit in his own country to make aggressive and intemperate assaults on the Catholic faith," and that it was "not merely bad taste but a departure from the methods of science for a naturalist to assail theological opinions." The *Herald* featured the lectures prominently, though it larded its reports with jocosities about "young ladies" who attended from fear of "having for the next ten days to reply negatively at the breakfast or dinner table to the query as to whether or not they had heard Professor Huxley." Of the final lecture, it scoffed that "anyone who casually dropped in would have imagined that Professor Huxley was demonstrator in anatomy in a college of veterinary surgeons," adding that he seemed to be "riding a hobbyhorse."

As usual, the *Herald*'s quirks reflected its owner's. Raised as a Catholic by his Irish mother, an immigrant, Commodore Bennett disliked the Church but disliked, even more, to hear others criticize it. Three months before he would urinate in his fiancée's parlor, "young ladies" must have weighed on his mind. The "veterinary surgeon" jeer reflected one of the rare subjects that aroused his sympathy. He was an animal lover, though he managed to carry even this to a repellent extreme. He became so fond of lap dogs that he never went anywhere without dozens. This led to his deep hatred of vivisection, with which Bennett associated such comparative anatomists as Huxley, who did defend experimentation on live animals.

Bennett seems to have decided that a domestic Darwinist was

preferable to a foreign one. The *Herald* presented the lectures as a triumph for Marsh, reporting that the second one had emphasized the Professor's toothed birds and the third his horses. "Of late years the marvelous tertiary deposits of the west of America had supplied fossils so numerous and in such perfect state of preservation that they left nothing wanting," the *Herald* paraphrased, "and the field itself had been worked by Professor Marsh, of New Haven, in a most incomparable fashion, so that nothing too much could be said of the care taken by him, the extent of the discoveries that had been made, or their scientific importance." Bennett's reporter noted that the audience "loudly applauded Professor Huxley's remarks touching on the eminent services rendered by Professor Marsh," and concluded that "the only thing wanting to complete the victory of *equus* over *fides* was one toe, which the Professor was confident would ultimately be found if they could only get at the geological formations in which the absconding little member had secreted itself."

The "absconding little member" referred to a fifth toe, which Marsh and Huxley believed would be found on a still undiscovered horse ancestor. They'd discussed the possibility of an earlier ancestor while going over the Peabody fossils, and Huxley playfully had drawn a five-toed cartoon and named it *Eohippus,* dawn horse. In his September 22 lecture, he predicted that such a genus would be found in the West. "Seldom has prophecy been sooner fulfilled," wrote his son. "Within two months, Professor Marsh had described a new genus of equine mammals, *Eohippus,* from the lowest Eocene deposits of the West, which corresponds very nearly to the description given above." Marsh didn't even have to go out and "get at the geological formations" to find his *Eohippus* (now known by Richard Owen's prior name, *Hyracotherium*). Its bones, with a vestigial fifth toe on each foreleg, were already in the Peabody. "I had him 'corralled' in the basement of our Museum when you were there," Marsh wrote to Huxley on July 12, 1877, "but he was so covered with Eocene mud, I did not know him from *Orohippus.* I promise you his grandfather in time for your next horse lecture if you will give me proper notice."

Huxley's sketch of a primeval man and horse

The *Herald*'s newfound admiration for Marsh the Darwinist did not extend to covering his *Eohippus* discovery, but that hardly detracted from his latest triumph. The Red Cloud affair had made him a national celebrity. Huxley's endorsement established him as America's leading evolutionist. "In accordance with your wish, I very willingly put into writing the substance of the opinion as to the importance of Prof. Marsh's collection of fossils which I expressed to you yesterday," the English naturalist had written to Clarence King just before presenting his New York lectures. "There is no collection of fossil vertebrata in existence which can be compared to it . . . It is of the highest importance to the progress of biological sciences that the publication of this evidence, accompanied by illustrations of such fulness as to enable palaeontologists to form their own judgement as to its value, should take place without delay."

When Marsh visited England in June 1878, Huxley flung wide the portals. "One evening in London at a grand annual reception of the Royal Academy, where celebrities of every rank were present," Othniel recalled, "Huxley said to me, 'When I was in America, you showed me every extinct animal that I had read about, or even dreamt of. Now if there is a single living lion in all Great Britain that you wish to see, I will show him to you in five minutes.' He kept his promise, and before the reception was over I had met many of the most noted men in England." A few years later, Darwin himself wrote to Marsh, "Your work on these old birds and on the many fossil animals of N. America has afforded the best support to the theory of evolution which has appeared within the last twenty years."

Despite residual anti-Darwinism among older academics, Othniel's new bishopric was a position of overwhelming strength. He consolidated his Huxley lecture triumph in 1877 with a famous keynote speech on North American fossil vertebrates at the American Association for the Advancement of Science's annual meeting in Nashville. "To doubt evolution today is to doubt science, and science is only another name for truth," he proclaimed. ". . . The classes of Birds and Reptiles, as now living, are separated by a gulf so profound that a few years since it was cited as the most important break in the animal series, and one which that doctrine could not bridge over. Since then, as Huxley has clearly shown, the gap has been virtually filled by the discovery of bird-like Reptiles and reptilian Birds . . . the stepping stones by which the evolutionist of today leads the doubting brother across the shallow remnant of the gulf, once thought impassable."

The *Nashville Daily American* printed the speech verbatim, and in an editorial called Marsh "a bold discoverer, whose services to the science of paleontology in our western wilderness have been more heroic and not less important than those of Cuvier," and his work "hardly second in importance to that of Darwin." Such enthusiasm might seem surprising in the Bible Belt, but Darwin's Church Scientific had long arms. The editorial's probable author, Albert Roberts, was an in-law of Huxley's sister, who lived in Nashville. Huxley had visited her after his Yale sojourn and made a speech on Tennessee geology that, Schuchert wrote, "was listened to with deep interest."

All this must have been hard to swallow for Cope, who attended the Nashville meeting. He'd worked as hard as Marsh. Why wasn't he getting attention? Edward didn't seem to realize that there was more to scientific success in 1877 than scientific work. "Both Cope and Marsh saw that the key to understanding the evolution of horses lay in the Eocene fossils from Wyoming and New Mexico," wrote Url Lanham, "but only Marsh was able to convert this insight into a commodity of real educational and publicity value. Cope's brief papers . . . failed to point up the drama inherent in the subject. The casual reader would not gather from his terminology that horses were even present in older Tertiary times, since he referred the earlier forms to other families. His technical diagrams of the phylogeny of the group were uninspiring."

Cope's prompt publication of new discoveries always had been frustrating for his rival, but the promptness turned to Edward's disadvantage as Marsh began to publish a series of long, carefully wrought monographs. Cope just kept churning out dozens of reports a year, and "this hurried publication led him to many errors of interpretation and nomenclature," his friend Osborn observed, whereas "the brevity and clarity of Marsh's descriptions and their uniform appearance in the *American Journal of Science* led to the immediate reception of his works abroad." Marsh's reputation began to grow "very much more rapidly than that of Cope, and his influence began to be felt in the geological surveys and in scientific academies abroad."

Cope's own restless brilliance hobbled him in the race for recognition, and this worked at a deeper level than publishing. He never could embrace the Darwinism that united Huxley, Marsh, and other mid-Victorian evolutionists. This heresy proved more damning than hasty publications, because it was based on scientific insight as well as emotional resistance to Darwinist agnosticism. Cope accepted that "the law of natural selection [is] the cause of modification in descent," as he wrote in an 1871 essay, "On the Method of Creation of Organic Forms." Yet he saw a basic biological problem with Darwinism. "This law has been epitomized by [Herbert]

Spencer as 'the survival of the fittest,'" he continued. "This neat expression no doubt covers the case, but it leaves the origin of the fittest entirely untouched." Cope had found nineteenth-century evolutionism's Achilles' heel. It didn't explain *why* organisms vary, which they must do, as he wrote, "in order that materials for the exercise of natural selection should exist. Darwin and [Alfred Russel] Wallace's law is, then, only restrictive, directive, conservative, or destructive of something already created. I propose to seek for the originative laws by which these subjects are furnished—in other words, for the causes of the origin of the fittest."

Cope's proposal was admirable, and Stephen Jay Gould, a paleontologist now at Harvard, has called him "America's first great evolutionary theorist." Edward sought the secret of variation in theories loosely called "neo-Lamarckian," building on the French naturalist's idea of the inheritance of acquired characteristics. He posited a "force" of will or consciousness that impelled organisms to develop traits of enhanced fitness through their behavior, a process he called "kinetogenesis." A process called "recapitulation" then would incorporate the new traits into embryonic development thereby passing them to the next generation. This was respectable science in the late nineteenth century, and Cope tried to show how the mechanical stresses of kinetogenesis might have caused evolutionary changes in fossil bones. Yet the search for variation's secret would lead down another path, one that Gregor Mendel was just beginning to follow in his monastery garden. It was the path of mutation, changes within the cell, and Cope's generation didn't know enough about cells to follow it.

Neo-Lamarckism had a number of other adherents in the 1870s and 1880s, but most contemporaries took Cope's theories less seriously than they took his other work. Darwin mentioned them in the sixth edition of *The Origin of Species,* but found them obscure. "It has quite annoyed me that I do not clearly understand yours and Professor Cope's views," he wrote in an 1872 letter to the neo-Lamarckian Alpheus Hyatt, "and the fault lies in some slight degree, I think, with Professor Cope, who does not write very clearly."

Osborn later recalled, "It was in violent contrast to the generally sound and well-based paleontological observations of Cope, that, like his great predecessor Lamarck, he should have been led into a tissue of hypotheses . . . In these studies and, in fact, all his Evolution generalizations, Cope ceased to be an observer, much less an experimentalist . . . always citing from observations and experiences not his own and accepting the baldest examples of supposed inheritance effects."

It wasn't as though the Darwinians had an answer to Cope's objection. In fact, as Björn Kurtén observed, "it caused Darwin himself reluctantly to include inheritance of acquired characters as part of the mechanism." Darwin tried to explain this "inheritance" in terms little sounder than Cope's; he concocted a theory called "pangenesis," which posited vague "gemmules" as units capable of passing on acquired characteristics. Darwin was Darwin, however, and Cope was not, as Edward was made to understand in 1878, when he visited Europe at the same time Huxley was introducing Marsh to the British lions.

"On some accounts it is not convenient to go," he wrote to his wife, "but I must counteract the Marsh mud, which appears to have stuck over there in several places." In his letters home, he tried to transmit his usual self-assurance. He "very much enjoyed" a Dublin meeting of the British Association for the Advancement of Science, "and made very many good friends." In Paris, the mostly anti-Darwinist French paleontologists nominated him to their Geological Society. Yet a defensive note kept creeping in. In Paris, "one person of importance" told him that "the criticisms of a certain person *'ne fait rien,'*" which must not have been that reassuring. In London, he had to explain to the editor of the three leading scientific periodicals "the matter of the dates of my papers, about which he has been a good deal fuddled by Marsh."

There was one undeniable indication that "the criticisms of a certain person" counted. "The only traces of Marsh's handiwork I could discover," Edward told Annie after the British Association meeting, "was in the indifference of Huxley, which I ascribed to that source. He took no pains to see me nor hear any of my papers —a coolness I suspect he would not have shown to friend Marsh." There's no record of Marsh running down his rival to Huxley, but, then, Othniel was a master of the subtle slight ("The energy of Cope has brought to notice many strange new forms, and greatly enlarged our literature," he'd purred at Nashville), and he wasn't the only establishment insider poisoning the episcopal ear. Louis Agassiz's biologist son Alexander had written to Huxley in 1874 that Cope had earned "the contempt of all the scientific men of the country."

Edward tried to put a bold face on Huxley's coldness. "I however introduced myself and I think we should have had some pleasant conversation," he wrote to Annie. But it was clearly disappointing to be snubbed by the man who, he believed, had foreseen a decade earlier that "the discoverer of *Laelaps*" might be the one to "elucidate" American paleontology. "Huxley has such influence," the letter continued wistfully, "not only on account of his writings but because of his readiness as a speaker and his quick wit, which I had occasion to hear several times."

10

Dinosaurs and Fate

Cope and Marsh had been squabbling for the better part of a decade by 1877. Cope's letters show how much it distressed him, and even the hard-shelled Marsh emitted an occasional cry of anguish. One day Samuel Williston happened to be in the Peabody laboratory, unnoticed, when the Professor finished reading a paper by Cope. "Gad!" Marsh had cried, inspecting a tray of fossils with the paper in hand. "Gad! *Gad! Godamnit! I wish the Lord would take him!*"

Professional rivalries seem inescapable, but few keep mounting as Cope's and Marsh's did. People turn away from the stress in various ways—carving their subjects into manageable specializations, agreeing to disagree. Had the two paleontologists tended in that sensible direction after the Fort Bridger debacle, both would have had brilliant careers just in the subjects they addressed between 1868 and 1876. Archbishop Huxley might have anathematized Cope, but there was room in nineteenth-century science for wealthy heretics. Edward and Othniel, however, proved incapable of resisting the peculiarly American temptations that confronted them. There was so much out there in the unexplored West, so many unmined minerals, unplowed soils, uncut forests . . . undug fossils.

"Marsh was never to have fossils 'enough,'" wrote Charles Schuchert. "Even with almost every box opened at the Museum yielding a new species, or throwing new light on an old one, he saw need for more and yet more material." Cope was the same. During his 1878 European trip, he paid $2500 for a collection of Argentinean mammal fossils that proved to have immense significance when the American Museum of Natural History acquired them in 1897, since they were mainly of groups peculiar to South America. Yet Cope never found the time to describe them. "In fact," wrote Henry Fairfield Osborn, "the boxes containing the Pampean collection were only partly opened, and, with other boxes arriving from various parts of the United States, remained unopened to the very end of his life. Cope himself never knew all the richness and variety which these boxes included."

From greed arose selfishness. Osborn and his friend W. B. Scott had become interested in paleontology from studying at Princeton with Cope's associate A. H. Guyot. When they decided to go fossil-hunting after graduation in 1877, they asked Cope for advice, probably expecting the avuncular encouragement they'd received in college. "We could extract no information from him," recalled Scott; "he was polite and pleasant enough, but absolutely noncommittal . . . When I asked him whether the country around Fort Wallace was good collecting ground, he answered: 'It was when I was there' and declined to say whether it still was, or not." The Princetonians instead went to Fort Bridger, where the Marsh forces gave them the same treatment they'd given Cope in 1872, to the extent that they actually became partisans of the unhelpful Cope on their return east. Of Marsh, Scott wrote, "I came closer to hating him than any other human being that I have known, and his hostility to me had a really detrimental effect on my career."

Such hubris virtually guaranteed that Marsh and Cope would collide again in the Western fossil fields despite any reluctance caused by past sufferings. Yet the circumstances under which they did collide proved improbably dramatic, as though the fate they'd been tempting was ruled by a James Gordon Bennett deity demanding the maximum sensation. They absentmindedly, simultaneously, stumbled on the most spectacular fossils of their careers; indeed, the most spectacular that had been found anywhere.

In 1877, a geologically inclined English clergyman named Arthur Lakes was teaching school at Golden, Colorado, in the Rockies' eastern foothills. One day in March, as Lakes looked for fossil leaves in Mesozoic strata near the little town of Morrison west of Denver, he found a startlingly large vertebra embedded in sandstone. "It was so monstrous," he wrote in his journal, "that I could hardly believe my eyes." A well-informed, enterprising individual, Lakes reported the find in local newspapers and got in touch with scientific authorities. He sent a sketch to Marsh, who responded with an offer to identify the bone if Lakes would send it to him. Lakes agreed to do so after looking for more, but Marsh didn't respond immediately, which apparently motivated Lakes to write to Cope. In any case, he sent bone samples to both men. Before the fossils reached him, Marsh roused himself to send Lakes a $100 retainer, along with his usual exhortation to keep the find quiet. Lakes answered that he appreciated the money but had already informed the newspapers and Cope about the fossils.

"It is not difficult," wrote Schuchert, "to picture the lively scene that Professor Marsh's office presented the morning this letter arrived." Gad! Marsh telegraphed Benjamin Mudge to drop everything and rush to Morrison to hold the site. Mudge arrived there June 28, and a few days later wired Yale: "Satisfactory arrangements made for two months. Jones cannot interfere." "Jones," of course, stood for Cope, part of a code Marsh had devised for communicating with his fieldworkers. "Ammunition," stood for money; "health" for success in collecting. Even more cryptically, the code word for pterodactyl was "drag."

Lakes's fossils were worth the cloak-and-dagger fuss. Mudge rightly called them "the very largest bones of dinosaurians or of any other saurians" he'd seen. They belonged to the largest dinosaurs —indeed, the largest land animals—that ever lived: the sauropods. Some species measured over a hundred feet from their disproportionately small heads to the ends of their whiplike tails. They were abundant in the Jurassic Period, from 200 million to 140 million years ago, and left thousands of bones in formations of that age in Colorado, Wyoming, and Utah. It seems surprising that the sauropods had caused no excitement before 1877, but perhaps paleontologists had lacked an intellectual context for such improbably large fossils. If the Cretaceous Period that Cope had sampled in the Judith River badlands the previous year was like an inky subbasement to the Age of Mammals, the Jurassic *below* the Cretaceous must have seemed another dimension.

Greed and selfishness now provided a broad context for excitement and for squabbling. Marsh rushed a "Notice of a New and Gigantic Dinosaur" into the July 1 *American Journal of Science,* estimating the animal to have been between fifty and sixty feet long, bigger "than any land animal hitherto discovered." He named it *Titanosaurus montanus,* which was apt, except that the name already had been applied to a dinosaur genus in India, as Cope promptly noted in the *Proceedings of the American Philosophical Society*. Edward's promptness is understandable, given that Marsh had shown, a few months earlier, that *Laelaps,* Cope's name for the dinosaur he'd discovered in Haddonfield, had already been applied to a spider genus. Marsh had renamed it *Dryptosaurus* ("wounding reptile"), but Cope ignored the change and kept using *Laelaps.* Perhaps to forestall Cope's doing the same to him, Marsh quickly changed his *Titanosaurus* to *Atlantosaurus.*

Yet it appeared that Othniel might not have to squabble over possession and priority of the Morrison fossils; Cope proved surprisingly accommodating about the specimens Lakes had sent him. Asked to surrender them to Marsh, he did so promptly, and withdrew from publication a paper he'd prepared, although he did read it before the American Philosophical Society. Marsh named the surrendered bones *Apatosaurus,* "deceptive reptile."

Edward's generosity was deceptive, in fact; he had a card up his sleeve. Another Colorado schoolteacher and amateur naturalist had found giant saurian bones in March 1877. O. W. Lucas had stumbled on them while botanizing in the Garden Park area near Canyon City, also part of the Jurassic Morrison Formation, and informed Cope, who asked for specimens. It's not clear when he received them, but in light of his whirlwind writing habits, it probably was only shortly before his article about them appeared on August 23. This must have caused more liveliness in New Haven, because Lucas's specimens included not just the jaw of a huge carnivore, which Cope happily assigned to *Laelaps,* but some beautifully preserved sauropod bones, as well. "What the total dimensions of this saurian are (*Camarasaurus*), is not readily estimated without further data," Edward began modestly. He quickly worked up a more characteristic head of steam, however, gloating that Lucas was sending him a six-foot-long femur, and finally bursting out: "This remarkable creature . . . exceeds in its proportions any other land animal hitherto discovered, including the one found near Golden City by Professor Lakes."

Marsh telegraphed Mudge to drop everything and rush to Canyon City, whence the former professor reported not only that Cope had already bought the bones "at cost, a low sum," but that Marsh could have had them if he'd paid attention to letters that the sourdough bone sharp David Baldwin had been sending him since the winter before. "There are bones of large animals in the Jura near here," Baldwin had written, "but I did not dig them on account of not hearing from you." Mudge examined Lucas's bones and told Marsh that they were "on a larger scale than *Titanosaurus montanus* from 10 to 30 percent," and were better preserved than the Morrison bones, because they'd been embedded in mudstone or shale, not hard sandstone. The sandstone near Morrison is so hard that big bones remain in the ground today, the only such "fossil feud" site I saw. Arthur Lakes had had to use dynamite to get his bones.

Mudge held out shreds of hope at Garden Park. He'd found more bones a couple of miles from Lucas's quarry, and Lucas was having doubts about his deal with Cope. "Secure all possible," Marsh telegraphed. "Jones has violated all agreements." Mudge settled down to dig at his sites, hoping to enlist Lucas after his contract with Cope expired. Things went badly, however. Mudge's site was on sandstone, not clay, and the bones crumbled when they were removed from the matrix. He asked Marsh to send Samuel Williston to help, and when Williston arrived Mudge went east, evidently fed up with giant saurians. Williston had the same problems, but, still a Marsh enthusiast, he wrote, "I don't propose to see Cope get better specimens . . . than we can." Marsh, however, soon ordered him back to Morrison. There, the small quarry he'd been working collapsed, almost killing his assistants and ending the digging season.

An exasperating interval, yet the wily Marsh also had a card up *his* sleeve. On July 19, he'd received a letter from two "Obedient Servants" named Harlow and Edwards, in Laramie, announcing the discovery "not far from this place" of giant bones and offering to send samples. "We would be pleased to hear from you," they wrote, "as you are well known as an enthusiastic geologist and a man of means, both of which we are desirous of finding, more especially the latter." The Professor doubtless received many such naïvely avaricious inquiries, but the bones this one described must have made him sit up. "We measured one shoulder blade and found it to measure four feet eight inches (4 ft. 8 in.) in length," the letter said. "One joint of the vertebrae measures two feet and one half (2½) in circumference and ten inches (10) in length." The "Servants" thought the bones were those of a *Megatherium,* Cuvier's Tertiary ground sloth, but no known land mammal ever had shoulder blades or vertebrae that size.

When the bones reached New Haven, on September 19, Marsh immediately telegraphed Harlow and Edwards to send more. He also sent a check for $75, but they answered that they couldn't cash it, which must have been irritating as well as mysterious. Marsh roused an ailing Samuel Williston out of seasonal retirement in Kansas and sent him to Wyoming. (Williston pretended he was going to Oregon in order to fool Charles Sternberg, who was snooping for Cope.) Arriving on November 14 at the Union Pacific whistle stop of Como, sixty-five miles northwest of Laramie, Williston at first could learn only that Harlow and Edwards lived at a distant ranch. "A freshly opened box of cigars, however, helped clear things up." It turned out that the Laramie correspondents had been so determined to keep their *Megatherium* a secret from "plenty of men looking for such things" that they'd used aliases. They were the Como station agent and section foreman, William Edwards Carlin and William Harlow Reed.

The secretive pair showed Williston the fossils at the nearby bluff, and he was awed and astonished that nobody had remarked on them before. The Hayden and King surveys both had mapped the area, and although Marsh later wrote that he had acquired large vertebrae during his 1868 salamander-collecting visit to nearby Como Lake, he clearly hadn't done much about it. Williston wrote to Marsh, on the day he arrived, that the bones extended "for *seven* miles and are by the ton." Bones lay right beside the station, "very thick, well preserved, and easy to get out." He promised to "send a ton a week gotten out good." The fossils were in soft clay and shale and, though miles from the station, could be brought to the tracks on wheelbarrows.

"Canon City and Morrison are simply nowhere in comparison with this locality both as regards perfection, accessibility, and quantity," Williston concluded in a November 16 letter, but added, "There will be great danger next summer of competition." Indeed, they learned a few weeks later that a miner named Brown had written to the Smithsonian about the fossils earlier that year, that the Smithsonian had informed Cope, and that Cope had asked Brown for samples. Marsh hastened to sign a contract stipulating that Carlin and Reed would take "precautions to keep all other collectors not authorized by Prof. Marsh out of the region." Although the Cope threat made him pay more than he wanted to, it was a wise move. The Como Bluff field turned out to be what Schuchert called "the greatest the world has so far known," and it immediately yielded rich results. Williston returned to Kansas for the winter, but Reed proved a persevering collector and sent carloads of bones east during the rest of 1877. Marsh was able to describe and name some of the world's most famous dinosaurs in the December issue of the *American Journal of Science,* including the spike-tailed *Stegosaurus,* the giant predator *Allosaurus,* and the sauropods *Diplodocus* and *Brachiosaurus.*

Como yielded something else that would burnish Othniel's evolutionist luster almost as much as toothed birds. In early 1878, Reed found a tiny jaw that had belonged to a mammal contemporaneous with the dinosaurs. A few such fossils already were known from the Old World, where they'd caused much debate, since it seemed incredible that mammals had lived so long ago. Marsh ordered his men to go through every shovelful of what they began calling "pay dirt," and they soon found more such jaws and teeth. "The cost involved in getting these tiny fossils makes their value greater than their weight in gold," Marsh characteristically remarked, and he smelted his paleontological ore into yet another nail for Cuvierian anti-evolutionism's coffin; he made a new order for them, the Pantotheria, the basic beasts. "The generalized members of this order," he wrote, "were doubtless the forms from which the modern specialized Insectivores and Marsupials, at least, were derived." So much for Cope's parvenu Creodonts.

Como Bluff's wealth of fossils has made it one of the world's most famous sites, but when I was there, I saw why King and Hayden had missed it. On my drive northwest from Laramie with the paleontologist Brent Breithaupt, I kept looking for badlands, but I saw only long swells of prairie, with the snowy Rockies in the distance. It was unusually wild prairie—little pronghorn herds were dotted about—perhaps virtually what the bonehunters saw, though it lacked the Gothic spires of Ferdinand Hayden's "cemetary." There was still no sign of a bluff as Breithaupt directed me onto a gravel road, just gently rising land spangled with white phlox, pink locoweed, and tiny yellow violets. A near gale-force wind shook the grass on that sunny day. "This is one of the windiest places in Wyoming," Breithaupt said. "People went crazy from it."

I saw no bluff as I parked the car on a rise, with the valley of the

Medicine Bow River to the west. But when we walked to the edge, the gray mudstones of the Morrison Formation became visible in low cliffs extending to the southwest. It was much bigger than it first appeared, although not much different from hundreds of other bluffs I'd driven past. Its most striking feature was its bareness. The only trees visible were some willows along the distant stream. Breithaupt pointed west, to where the Union Pacific tracks once ran and Como Station once stood. "It was just a water and maintenance station, and it was abandoned in the late 1890s," he said. "There's less civilization here now than there was then."

We walked down the slope to an eroded pit at the bluff's base. "This is the famous Quarry Nine, where they've found more Jurassic mammals than anywhere else in the world," Breithaupt told me. Mammal fossils weren't in evidence, but there was no doubt the dinosaurs had been there. Entering his paleontologist trance, Breithaupt quickly began picking up their striated, blue-black bone fragments. He gestured at the ashy mudstone. "It's lucky it hasn't rained much lately. When this stuff's wet, you can hardly move; it sticks to you in layers. Then when it dries out later in the summer, you get sandblasted. After a day out here, I'll come away with my gums packed with grit." I expressed awe at the 1870s collectors' endurance, and Breithaupt nodded, adding that conditions weren't all that different then. "The winds usually blow from west to east here, so they could have got out of it with tents. They'd put up tents over the diggings as well as to live in. They could be pretty comfortable." He also told me that trains to Wyoming in the 1870s had included Pullman sleepers. "Cope and Marsh probably rode in a fair amount of style," he said.

It was the social rather than the physical environment that caused the worst discomforts for the Como Bluff diggers, who seem like characters in a Gilded Age *Treasure of the Sierra Madre.* Of course, the bluff's richness piqued Cope's greed, and he had learned from experience about employing field hands for spying and dirty tricks. Marsh's attempts to pre-empt the site would have been like a red flag to him. Treachery became a way of life as bone sharps arranged and rearranged themselves like iron filings around the magnetic poles of professorial greed and selfishness, a story that Marsh's correspondence with his collectors documented in uncharacteristic detail.

The treachery began early, as the first bone shipments went east. After negotiating the contract with Marsh, Carlin let Reed do most of the backbreaking work of getting the fossils from bluff to railroad, and he may have leaked the find to the *Laramie Daily Sentinel*, which published an article about it in April 1878. Calling the fossils "giant crocodiles," the piece exaggerated what Marsh had paid for them, as though its source hoped to jack up prices. Carlin, who complained early and often about Marsh's payment delays, sent a clipping of the article to Yale, along with a letter: "It would be well to hasten operations as much as possible, as it will probably be included in the Associated Press report . . . and it may be difficult to keep the other parties out."

Marsh tried to plug the leak by sending out Williston, who reported that a suspicious character, a "thick heavy-set sullen portly man" calling himself Haines, was asking about fossils. "There is no doubt that he is direct from Cope," Williston wrote. "Carlin at the time thought he was Cope, and tried very hard to catch him. The chief reason that I *know* it is not he is that the man wrote a very neat, fine, and legible hand and I have seen Cope's writing! . . . Didn't mention Cope's name, but yours frequently, *rather* disparagingly! When Carlin called Cope a 'damned thief,' sneered . . . If you recognize him let me know as soon as possible and I will have a little fun at his expense." Haines departed unidentified, however, and Reed and Williston's brother Frank worked diligently through the summer. Williston himself had become disaffected with Marsh, so he spent most of his 1878 Como sojourn collecting birds and spiders for his own edification. In August, the team dispersed for the winter, though Reed had to stay in Como a month waiting to get paid. He bore the delay stoically; Carlin threatened to sue Marsh for the money.

Treachery incubated during the long Wyoming winter. There

evidently had been more than met the eye to Carlin's ostentatious hostility toward "Haines," because when Reed returned to collecting in February 1879, he found that his erstwhile partner had gone to work for Cope. The stationmaster, Carlin, would not even let Reed use the freight room to box his specimens, so he had to do it on the wind-whipped platform and camp out at a quarry. "It is just merely H———, H———, H——— . . . ," he wrote to Samuel Williston in New Haven. "They are doing lots of talking down at the Station, but talk does not hurt me nor scare me much either." Two strangers invaded one of his quarries, obviously hunting for fossils, and Reed repelled them by climbing to the bluff top and flinging down dirt and rocks on them for several days. On at least one occasion, he smashed bones that he didn't collect so that Carlin and Cope couldn't have them. Informed of the situation by Williston, Marsh ordered Arthur Lakes to stop work at Morrison and help Reed.

Hard-bitten Como fascinated and unnerved the genteel Lakes. "The station consisted of a red building, a tank like a huge coffeepot and a small red section house," he wrote in his journal. "Bidding the Bishop good bye, I jumped onto the platform and walked into the station house room. There were half a dozen men smoking and playing cards around the stove. Amongst them was a tall swarthy complexioned man with a handsome face dressed in a full suit of buckskin ornamented with fringes after the frontier fashion . . . the most striking and commanding member of the group looking like the ideal of a frank, freehearted frontier hunter." The picturesque figure was Reed, who regaled the Englishman with stories of desperadoes and shootings, including a recent one about bandits with a hideout on Como Bluff. They'd tried to wreck the train and then killed two sheriff's deputies. As one captured bandit was being taken to his trial, Reed said, "the miners entered the train, took out the prisoner, hung him to a telegraph pole, and riddled his body with bullets."

Reed was ready with a gun himself; he blasted antelope, deer, and elk for meat and other animals for fun, including a small dog whose barking kept him awake one night. Lakes also took to blasting things—there wasn't much else to do for entertainment. "The trains constantly passing and repassing keep us in mind of our being after all within reach of civilization," he commented. "It seems curious however from our wild life to look in on the luxurious cars and see well-dressed gentlemen and ladies like phantoms we once knew passing by from some other world."

Marsh visited on June 4, soon after Lakes's arrival, and together they explored the bluff from a railroad handcar. "It was a lovely morning," Lakes wrote, "and Professor M was much amused at the innumerable spermophiles, rabbits and hares that kept leaping in every direction from the railway track as we sped along. We stopped first at Quarry No. 3. Professor M, after examining carefully the bones, devolved many of them to destruction as too imperfect or rotten for preservation." They fossil-hunted on various parts of the bluff, and after supper Marsh regaled them with anecdotes of the Fort Bridger expeditions.

Othniel was having so much fun that he decided to stay another day, during which they went to the train robbers' former hideout and lunched "under the identical cottonwoods that had sheltered the bandits." A talented if naïf artist, Lakes painted a watercolor of the derby-hatted Professor, Reed, and another bone sharp named Ashley, all in poses oddly suggestive of Manet's *Déjeuner sur L'Herbe.* Afterward, "Professor Marsh, full of excitement and enthusiasm leading the way," they discovered another fossil site containing a jaw "with small crenated teeth like those of the *Iguanadon* or the modern iguana, an herbivorous saurian." Evidently feeling he'd unearthed enough "with his own hands," Marsh departed the next morning.

In August, Cope dropped in, throwing Marsh's men "into a frenzy of patrol activity, trying to occupy all the quarries simultaneously," as Url Lanham put it. Marsh doubtless had embroidered the tale of Cope's villainy during his June visit. Edward's two days at Como may have fallen short of the Mephistophelian expectations he had raised, however. Arthur Lakes described him as "a tall, rather interesting-looking young man" who chatted about England and geology and "entertained his party by singing comic songs with a refrain at the end like the howl of a coyote." Cope's poor health may have contributed to this benign impression. "I felt that I was going to have one of my attacks of fever," he had written to his wife from a prior stop in Canyon City, "and yesterday I had it sure enough. I slept much and took aconite every hour day and night." He seemed to be disoriented by the vast changes in the West, with cattle now on former bison ranges.

"Today I made a grand exploration over the bluffs, where I had never been before," Edward wrote to his daughter, Julia, on August 8. "I found the men who have been working for me were not the ones with whom I had been corresponding, which quite surprised me." Apparently, he had been out of touch with the ever-unreliable Carlin, who had passed the bonehunting chores to a pair of brothers named Hubbell. "I saw bones of *Camarasaurus* sticking out of the ground," Edward continued excitedly, "and the boys have dug up a huge, flesh-eating saurian which they will send off in the morning."

Cope's letters may have been less than candid about intrigues likely to offend his Quaker wife's sense of propriety. As Jane Pierce Davidson observed, he liked to tell his family "what he felt they wanted to hear." Arthur Lakes expressed some ambiguity about the visitor in an August 11 report to New Haven: "The *Monstrum horrendum* Cope has been and gone, and I must say that what I saw of him I liked very much, his manner is so affable and his conversation very agreeable." But he added, "I only wish I could feel sure he had a sound reputation for honesty."

Marsh's men had a productive summer as they rushed about trying to beat Cope's, but they began trying to beat one another as well. Despite Lakes's admiration for him, Connecticut Yankee Reed disliked the Englishman's haughty Oxford manner, and complained to Williston that he was a malingerer. He regarded Lakes's paintings as a waste of time, although Marsh had requested them as a record of the site. Lakes was less vocal about the conflict, but his journal entries on the "freehearted frontier hunter" dwindled. In late August, both men offered their resignations, and Marsh separated them as if they were squabbling schoolboys, sending Lakes to the bluff's southwest end and Reed to the northeast.

"My work has been solitary, my only company being a few large hawks," Lakes wrote in September, adding apprehensively that "Reed and Cope's men have been preparing for the severities of the coming winter by making dug outs . . . I shall stand in my tent as long as I can and then take refuge in the section house." His unease increased in October, as formerly peaceful Utes killed a missionary and soldiers in the mountains to the west. "Troops kept passing through in the trains and with them all the equipment of war," he wrote. "On the 3rd the whole northwestern horizon was enveloped in smoke, leading us to suppose the Indians had fired the woods." Crowds of packrats and mice invaded his tent at night, as did an occasional swarm of tiger salamanders. One of the rodents' "nightly amusements," Lakes complained, was to fill his boots and shoes with beans and rice. They even tormented him at work. "On arriving at the quarry," he wrote on October 14, "I found my broom carried fifty yards from the hole down a steep ravine by mountain rats, as testified by their teeth marks. My butcher knife and ball of string were also missing."

The next day it "blew and snowed," but Lakes stuck it out through the fall and winter, immortalizing his hardships in a painting called "The Pleasures of Science," which shows a figure like an Eskimo clutching a pickaxe and squinting against wind-driven snow. Because the strata at his quarry lay at an almost vertical angle, he had to dig a virtual mineshaft to get at bones. "Collecting at this season is under many difficulties," he wrote on February 5. "At the bottom of a narrow pit 30 feet deep into which drift snow keeps blowing and fingers benumbed with cold from thermo between 20 and 30 below zero." When he got to forty feet, he hit a spring, and "had to bale [sic] with one hand and dig out bone as I got a glimpse of it with the other, sitting at the time in a frog pond more like fishing for eels than digging for bones meanwhile snowing and freezing hard." He finally had to abandon the quarry, because it was about to collapse.

Lakes gave up in March and returned to teaching, but his departure didn't leave Reed in peace. Marsh replaced Lakes with a man named Kennedy, who had worked well on the railroad as Reed's assistant, but whose new status went to his head. He quarreled with his former boss, which led to the resignations of both their assistants, Ashley and Brown (possibly the Brown who had informed Cope of the Como fossils in 1877). Marsh sent Samuel Williston's brother Frank to make peace, and visited the bluff again himself in September, but discord prevailed. Reed and his assistants ostracized Frank, who moved out of the Marsh camp to stay with Carlin at his nearby ranch and by November was working for Cope. Reed suggested that Marsh freeze out Cope by buying up the quarry sites under the Desert Lands Act, but processing the claims took most of the winter, and groups of up to six Cope men kept encroaching. They tormented Reed as the packrats had Lakes. He wrote to Marsh on December 5 that they'd been to a quarry the day before. "They did not come to the quarey but was all around it then went to those big Bones west of the quarey . . . I find their tracks all around no. 9."

In the spring of 1881, Reed went to the expedient of hiring his own brother, but this attempt to get a reliable assistant failed in July, when the brother broke his neck diving into Rock Creek. Reed began to despair. "This country is run over with bone hunters and have been trying to hire my men," he wrote to Marsh on September 1. "They offerd Phelps more than you are paying me but I told him they would not give him steady work and you would not hire him again." Professor Alexander Agassiz of Harvard (who'd run down Cope to Huxley in 1874) had been sending out representatives, and Carlin, Frank Williston, and a man named McDermott had formed a "Bone Companey" to sell fossils to the highest bidder. Reed reported that even Carlin had been outraged when Williston and McDermott, who "did not know a bone from a stone," colluded to sell fossils from the same quarry to both Cope and Agassiz.

The disgusted Reed finally quit and went into sheepherding in 1883, and without his steadying influence the bluff degenerated into True West melodrama. Marsh, never having learned about chains of command, hired Brown independently of Kennedy, whom Reed had left in charge. This new rivalry finally boiled over in December 1884, when Kennedy wrote to Marsh that "a man working for you by the name of Brown assaulted me at station with two revolvers and wanted me to fight him. i refused on account of my family and he said to agents wife before i got thear he would shoot me . . . he must not molest me or i will put him under Bonds to keep the Peace." Despite his protestations, Kennedy was intimidated, and he left the bluff a few months later. The quarries from then on yielded little for Marsh, but he kept the trigger-happy Brown on until 1889, partly in hope of finding more mammals, partly to make sure nobody else found anything.

The bone war had moved far beyond Como Bluff by the time a windy peace descended, but it had reached a point of no return. If withdrawing from competition had been an option, it wasn't after 1878. For all the clamor of the bone sharps' treacheries, the most significant thing about the Como story was the stealthy persistence with which their wealthy masters tried to subvert one another. It was as much the fear of an antagonist's finding new fossils as the desire to possess them that was the driving force. Rivalry had begun to overshadow the paleontology that had initiated it and acquire a life of its own. Competition has a way of devouring any real interest in matters apart from itself, and dinosaurs came to seem almost beside the point.

Como Bluff retains a monument to the scientific hubris of fighting over giant saurians. The last place Brent Breithaupt showed me there was Quarry 10, a squarish pit just south of the mammal quarry. Quarry 10 is famous as the place where Reed and his assistants found the bluff's largest and finest quantities of sauropod bones, but it also is the site of history's most famous sauropod mistake. "In 1883, Marsh rushed to name the genus *Brontosaurus* from bones found here," Breithaupt said, "also using head, tail, and

Arthur Lakes's watercolor of the Como Bluff dig

some limb bones from other sites to complete the skeleton. *Brontosaurus* was the first dinosaur that Marsh restored, and it . . . created the standard image of the dinosaur that lasted a hundred years, and that we grew up with. But it was an animal that never existed. Marsh didn't realize that the Como bones belonged to *Apatosaurus,* a species he'd already named in Colorado, or that the bones from other sites that he used to complete the skeleton came from Cope's *Camarasaurus* . . . Marsh accidentally concocted a paleontological hybrid. He probably went to his grave unaware of the mistake, luckily for him. Cope would have been delighted to learn that Marsh had made a new genus by mixing *Camarasaurus* and *Apatosaurus,* but he never realized it either."

Cope and Marsh made major contributions to dinosaur studies. In 1968, the paleontologist Edwin Colbert wrote that dinosaurs had been "imperfectly seen" before their work, and that they had left them "well documented; described from complete fossil skeletons that gave a picture of the giants of yesteryear in which the elements of conjecture were reduced to a minimum." In the 1970s,

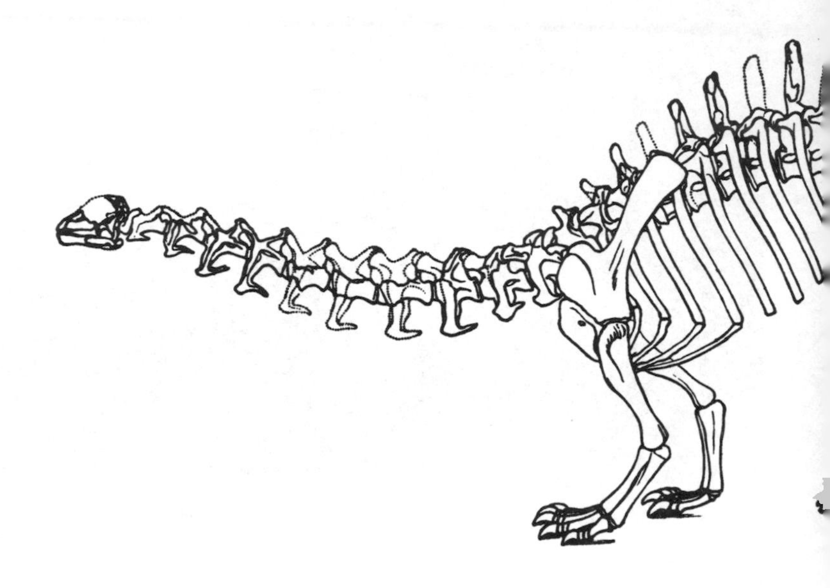

paleontological "young Turks" like Robert Bakker and Adrian Desmond saw the rivals as pioneer exponents of the "heretical" view that dinosaurs were not the sluggish reptilian creatures imagined in the 1950s but were lively birdlike ones.

Yet the twentieth century's dinosaur mystique has overemphasized and distorted those contributions. Cope and Marsh had "modern" ideas about dinosaurs not because they saw in them the mythologically compelling beings that obsess us, but because they were great paleontologists who analyzed bones with insight even when they did so in haste. The huge fossils really were more than either man could handle. "Marsh and Cope were almost literally buried by the tons of dinosaur bones . . . that were paraded before their eyes," admitted Colbert. "There was an embarrassment of riches, as a result of which they were unable to probe exhaustively into many of the problems presented by the paleontological riches surrounding them."

Marsh's Brontosaurus, *"a paleontological hybrid"*

Although Cope's nearly fourteen hundred published papers included many on dinosaurs, they were not as exhaustive as his mammal writings, and he did not study many of his specimens. He never knew that the "huge, flesh-eating saurian" the Hubbell brothers dug up during his 1879 Bluff visit is one of the most complete *Allosaurus* skeletons ever found. The boxes with its bones remained unopened until his death, and weren't even stored in his Pine Street laboratory. "This is mentioned as an example of the extreme pressure upon his time and means, and the fact that, like Marsh, he was collecting fossils more rapidly than he could possibly describe them," wrote Osborn, who eventually had the *Allosaurus* mounted for the American Museum of Natural History. A book of memorial addresses published by the American Philosophical Society after Cope's death lauded his achievements as a herpetologist, ichythologist, mammalogist, and geologist, but barely mentioned his work with dinosaurs.

Marsh did more in that area than Cope. He described nineteen still-valid dinosaur genera, in contrast to Cope's nine. He published fifty-five papers on them and made the first restorations of such famous ones as *Triceratops* and *Stegosaurus.* Yet this didn't match the evolutionary significance of his toothed-bird and horse studies, and his main writing in the 1870s and 1880s was on mammals. He planned a series of monographs on dinosaurs, but completed only a hundred-page summary, *Dinosaurs of North America.* Schuchert called it "probably his greatest work," but that was because it served as "the foundation of dinosaur knowledge." He acknowledged that it was "not based on as significant material from an evolutionary viewpoint" or "as finished a product" as the monograph on toothed birds. Marsh wrote that his dinosaur work had lasted "more than a score of years" and had led him across the Rockies "a still greater number of times," but he may have exaggerated. "In a total of three-and-a-half days," scoffed his main field assistant, John Bell Hatcher, "he seems to have found sufficient time to 'carefully explore' the geological deposits of the *Ceratops* beds and to trace them for eight hundred miles along the eastern flank of the Rocky Mountains, besides making numerous other observations of scientific interest." When, just before his death, Marsh listed the seven "prominent features" of scientific significance in his collections, the fossil horses and toothed birds came first. The dinosaurs came last, after uintatheres, brontotheres, pterodactyls, and mosasaurs.

Indeed, the two paleontologists' entanglement over dinosaurs may have worked against the main scientific enterprise they had embarked on in the 1860s. Their mammal studies vividly demonstrated evolutionary transitions and showed that such transitions had occurred throughout the recent Tertiary Period, thus linking mammalian humanity to the transmutational continuum. Their dinosaur work demonstrated few such transitions, and the bizarre Mesozoic Era was so distant from the present that it was easily seen, Cuvier-fashion, as unconnected with human life. As the fascination with dinosaurs grew in this century, many people came to regard *all* large, extinct animals as dinosaurs, a blurred vision of the evolutionary past as a never-never land with no link, and thus no relevance, to the present.

In their own time, the Como Bluff diggings were more like one of the more squalid gold strikes than an instance of exalted scientific discovery. Mean-spiritedness blew through them like the Wyoming winter wind, and the consequent intellectual poverty extended to the public reaction. This seems startling, now that Como Bluff dinosaurs are the bone war's best-known feature. In a 1992 PBS documentary, Robert Bakker stood before the bluff in his cowboy hat and sourdough's beard and declared that "suddenly dinosaurs became famous worldwide" after the 1877 discoveries. Since his innovative book, *The Dinosaur Heresies*, did so much to create the dinosaur mystique of the late twentieth century, Bakker's words have weight, as reflected in other recent writings on Cope and Marsh. The 1877 discoveries are at the center of most of these books and have even entered popular fiction. One science fiction story has Edward and Othniel rushing west to squabble over a living sauropod captured by a bunch of outlaws.

Giant bones did become a facet of Marsh's media personality. In 1890, *Punch* would caricature him as "Ringmaster Marsh," a pear-shaped figure dressed in the Stars and Stripes and standing on a *Triceratops* skull to put a pair of towering skeletons through circus paces. The performing skeletons in *Punch,* however, were those of a uintathere and a titanothere, Tertiary mammals, which accurately reflected contemporary public perceptions. Despite Como Bluff's wealth of fossils, its dinosaurs did not capture the attention of the Gilded Age. Not even the *Triceratops* skull in the *Punch* caricature came from Como Bluff's Jurassic deposits.

Except for local coverage like the *Laramie Daily Sentinel*'s "giant crocodiles" article, newspapers almost completely ignored the 1877 sauropod discoveries. I combed the *New York Times* and *Tribune* for days (as mind-numbing an occupation as combing dirt for bones), and turned up many pieces on evolutionary subjects—mammoths, fossil footprints, "paleolithic" human artifacts, Darwinism's growing influence. Even paleontological politics got some space. An April 28, 1877, *Tribune* article described Hayden, Cope, Marsh, and Powell arguing so "sharply" about the Cretaceous-Tertiary boundary at a National Academy of Sciences meeting that President Joseph Henry had to cut them off, "expressing his regret that the short time at the disposal of the Academy would not permit a continuance of this interesting debate."

I found one brief mention of sauropods. "Of the enormous reptiles of geology, none has yet surpassed in size one which Professor Marsh has recently described," the *Tribune* noted in an admiring August 10–11 feature on the Peabody Museum's opening. Yet the "enormous reptile" figured only briefly at the end of a detailed catalogue of the museum's other fossils, and the *Tribune* got its age wrong and made a joke of it. "Its remains were found in the Cretaceous beds in Colorado; it is named *Titanosaurus montanus;* it was herbivorous and attained a length of fifty or sixty feet," the feature offhandedly remarked. "Believers in a vegetarian diet should take note of these facts, as that saurian was the largest land animal known to have existed on this planet."

Both the *Tribune* and *Times* gave much more, if not more serious, space to a giant saurian supposedly still alive in the Mississippi River. On September 7, the *Times* carried an editorial on a barrel-bodied, snake-necked "Mississippi Monster," archly speculating that it was "a solitary ichthyosaurus which has survived the extinction of the rest of its species, and which has been completely renovated and lavishly decorated for the summer season on the Mississippi." An October 10 *Tribune* editorial was more informative (ichthyosaurs were dolphin-like) but equally jocose. It described the monster as a "pterodactyl-plesiosaur" with a twenty-foot-long neck, an enormous bill, immense fangs, and a mane of coarse red hair. A *Times* editorialist recycled the gag on October 20: "It is not every one who is able to discover a really meritorious monster," he wrote. "Had the person who first saw this complicated beast been content to make it an enormous lizard and to exhibit it as the sole survivor of the *Megalosaurus,* he would have deserved respectful attention; but he was so unwise as to give it the beak of a bird, the head of a bull, and a body composed of half a dozen other animals, thus bringing the monster into incredulous contempt."

The Mississippi monster was not the year's only evolutionary tall tale. On December 8, the *Tribune* reported a "Second Cardiff Giant," a seven-and-a-half foot, 600-pound "petrified man" that had been dug up in Colorado after its toes were seen protruding from the ground. The paper scoffed that John S. Newberry, professor of geology at Columbia, saw nothing genuine about the figure and that P. T. Barnum had been in the vicinity when it was found. (Barnum had exhibited his own Cardiff Giant in 1869.) The *Times* was equally skeptical, calling for public dissection of the figure. Yet skepticism didn't stop the *Tribune* from devoting several columns to the matter. While mocking Barnum's latest flimflam, a *Tribune* editorial just barely touched on the real discoveries the press was *not* reporting. "Wonderful fossils have been found in Colorado," it remarked, "and also much gold-bearing quartz. This is evidently not a fossil; perhaps, though, it may prove to be gold-bearing limestone."

House-size prehistoric saurians must have seemed right up James Gordon Bennett's alley. The *Herald* was much less dismissive of the season's tall tales than were the highbrow papers. "Professors Taylor and Paige pronounce that the figure that formed this mold was once a living human organism," it reported of the Colorado Giant, adding solemnly that "the greatest curiosity among the scientific men is the vertebra, which is extended about two inches and a half, displaying a well-defined tail." Considering the Onondaga Giant vexations of 1869, this apparent credulity may seem surprising. Yet Bennett was no slave to consistency if a story seemed marketable, although the *Herald* did *not* take notice six months later, when *Popular Science Monthly* reported not only that George Hull, the Onondaga Giant's creator, had contrived the Colorado one, but that "Professor Marsh was again called upon for an opinion, and at once detected the fraud."

The paper lavished space on the Mississippi monster, reprinting a *St. Louis Globe Democrat* article about an attack on a barge at a place called "Devil's elbow cut off," during which the monster supposedly left part of its beak in the vessel's side. "There are probably some who insist that there had been a bending of some elbows beside the elbow of the river . . . ," said an October 9 editorial, "but we see no reason to be skeptical . . . There are some queer inhabitants down in the deep waters." The *Herald* did not compare its monster to ichthyosaurs, plesiosaurs, or pterodactyls, however, and I found no mention of dinosaur discoveries in its frequent coverage of Western events, including three lengthy accounts of Hayden survey parties in August. A brief July article on the Peabody Museum's opening noted that Marsh had paid $1000 for a pterodactyl; it ignored his having just discovered and described the biggest fossils ever.

To be sure, Commodore Bennett suffered from distractions in that year of parlor urination. After his former fiancée's brother, Fred May, had horsewhipped him outside the Union Club until "blood stained the snow from sidewalk to gutter," as one paper put it, Bennett demanded satisfaction. Possibly the last formal duel in North America, their meeting, on January 7, 1877, took place at Slaughter's Gap, on the Maryland-Delaware border. Both men fired into the air, although there was some doubt whether Bennett meant to do so. The party then went to a nearby inn and ordered beer. "The proprietor of the hotel mistook his guests for pickpockets," reported the *New York Sun*, "and consequently sat up all night watching their movements."

Within a few months, Bennett expatriated himself to Paris, where he largely remained for the rest of his life, shuttling between châteaux, palaces, and yachts. In lieu of a wife, he acquired a mistress known only as Madame A., a Russian émigrée whom the Commodore was rumored to have chosen because he'd heard she was the most disagreeable woman in Paris society. His Slaughter's Gap brush with extinction had made an impression, however. Years later, when he heard Fred May was in France, he was rumored to have worn a coat of mail under his suit for weeks.

11

An Inside Job

The three-ring circus of the 1877 dinosaur discoveries wasn't even the sharpest Cope-Marsh clash of the period. After Marsh's triumph at the American Association's Nashville meeting that summer, Cope went to Texas and began discovering unknown fossil beds at an alarming rate. San Antonio's subtropical setting stimulated him. "This is almost part of the Mexican zoological region," he told his wife. "In the neighborhood are the Jaguar, the ocelot, also peccarri in herds, then the northern beasts Coyote, Antelopes, and Buffalo . . . There are great prairies between here and Houston . . . bounded by forests of *live oak,* evergreen and 'Bearded with moss' most beautifully. I wish you could see it." It must have seemed a prehistoric landscape come to life.

Jack-in-the-box quickly produced a major find, the Permian Period, which lasted for 50 million years and ended the Paleozoic Era about 230 million years ago. The discovery was less epochal than his Paleocene one, because the Permian already was well known in Europe, but nobody had found fossils of its giant amphibians and primitive reptiles in North America, and Cope didn't have to wait until 1881 to get them. A Swiss collector named Jacob Boll, whom he'd met in Dallas, found thirty-two species on the Texas plains in the winters of 1878 through 1880, and died in the field, "through his indifference to his personal comfort," as Cope regretfully reported in *The American Naturalist.* "Immediately," Osborn wrote, "Cope began his classic series of papers on the Permian vertebrates, first recognized in 1878 as 'a new fauna.'"

Cope's 1874 San Juan Basin discoveries had aroused Marsh's envy, and his Permian one drove the usually judicious Yale Professor around the bend. Marsh had declared in his Nashville speech that "no vertebrate remains" were known from the "Permian rocks of America," so the abrupt challenge to his authority must surely have disturbed him. That some of Cope's Permian fossils had traits that typified the transition from reptile to mammal probably did, too, since they implied an evolutionary link rivaling the toothed birds. Cope's Permian fossils would force Huxley to revise his concept of the reptile-to-mammal transition.

Marsh carried on as his more impetuous rival might have, and his behavior occasioned one of the worst accusations that would be leveled against him during the 1890 *Herald* flap. When asked about Marsh's professional ethics, W. B. Scott burst out: "Marsh distinguished himself by open theft some years ago. He listened to an address by Professor Cope before the National Academy of Sciences on the Permian reptiles. He took notes, and as soon as the address was completed hurried back to New Haven and held back the *Journal of Science* until he could print and bring out the discoveries as his own. In science, priority of discovery is secured by first publication. It is in this manner that Professor Marsh has won his laurels as a scientist."

Scott made the statement off the record, but the *Herald* published it in its bone war coverage, inciting Marsh's fury. "There is no truth in this statement," Marsh wrote to the *Herald.* "I had for years been making a collection of Permian fossils from the Southwest, and had already secured one of the largest collections made in this country. These I had long been investigating and most valuable results had already been attained. Many of these were ready for publication, and before I went to the meeting of the Academy, I had a short paper in print on that subject. I learned nothing from

Professor Cope or his paper except that a field which I supposed to be my own had been invaded in a manner not especially creditable to him. My paper was published on my return, with no changes of any importance." Marsh's *Journal of Science* paper, published in May 1878, ignored Cope's National Academy address and said that "hitherto no Permian vertebrates have been identified in this country."

Cope himself reiterated Scott's accusation in the next day's *Herald,* however. "After I read a paper on the Permian vertebrates before the National Academy," he told a reporter, "Professor Marsh left the presidential chair and disappeared from the meeting. In a few days a paper on the same subject by Professor Marsh appeared, in which some of the same discoveries were recited without credit, but with changed names. As I had previously fortified myself by a publication in *The American Naturalist,* the attempted plagiarism fell flat."

The *Rashomon* quality of this event seems to defy objective judgment. Yet Marsh's 1878 paper did not support his claim of lengthy investigations; it consisted of a few descriptive paragraphs and bone measurements on four supposedly reptilian species, and was unillustrated, though he usually made a point of using pictures. "His work of identification was so superficial," observed Professor Alfred S. Romer of Harvard, "that of two species described as belonging to a single genus, one is an amphibian, the other a mammal-like reptile!" Romer speculated that Marsh had owned Permian fossils in 1878, but that they'd been sitting in unopened boxes since David Baldwin had sent them from New Mexico years earlier. Perhaps recalling the consequences of his ignoring Baldwin's 1877 Canyon City dinosaur letters, the Professor may have indulged in a frenzy of box-jimmying and bone-describing on his hasty return from listening to Cope's paper. *"Gad! Godamnit!"*

Marsh would not have felt much like Huxley's magician as he fumbled through his Permian crates. It must have seemed to him that Edward Jack-in-the-box would be jumping up at him forever. And perhaps that would not have been a bad thing. Had the rivalry continued with any degree of equality, both men might have developed some tolerance. But fate already had placed in Marsh's hands an almost magical power to suppress such vexations. As Cope mentioned in the *Herald,* Marsh occupied the National Academy's "presidential chair" in 1878. He wasn't elected; he'd stepped into the position in April, when the sitting president, Joseph Henry, died. But he was the acting officer when a crucial matter arose later that year.

Congress long had been dissatisfied with the inefficiency and expense of having several geological surveys in the West, each operating independently. The Hayden and Wheeler surveys had mapped parts of Colorado twice and given some mountains two names. Although the Grant administration said it would consolidate the surveys under the army, it had done nothing. The question, therefore, came up again when Rutherford B. Hayes's new secretary of the interior, Carl Schurz, set out to reform public land administration in 1877. John Wesley Powell had been one of the spurs of congressional discontent. He lobbied for consolidation by presenting the reasonable argument that an efficient, scientifically sound survey was needed for orderly settlement of the West. According to Wallace Stegner, Powell saw an opportunity to put his association with Marsh to good use. The Yale Professor "was more than an illustrious scientist with a firsthand knowledge of the West; he was a man of power, shrewd in political manipulation, solidly backed"—and he was president of the National Academy of Sciences.

In the summer of 1878, probably at Powell's suggestion, Congress sought advice about the surveys from the academy. When Marsh returned from his Huxley-led tour of British lions in September, he appointed a six-man committee to draft recommendations on the consolidation issue, with himself as ex officio chairman. Packed with Marsh allies, the committee was highly receptive to Powell's point of view, and drew heavily on his just-published *Report on the Lands of the Arid Region of the United States,* which advocated a consolidated survey. Powell spoke at meetings and otherwise acted like an unofficial member, so it was not surprising that the committee drafted recommendations along his lines. "It was a tight inside job," Stegner observed. Powell's confidential clerk, James Pilling, wrote to his boss that the academy report sounded "wonderfully like something I have read—and perhaps written—before." So tight was the job that the academy adopted the committee's draft recommendations with a single dissenting vote—Cope's—and sent them to Congress in November.

Despite opposition from most Western congressmen, the House of Representatives passed an act consolidating the surveys in March 1879. Hayden maneuvered to become head of the new survey, and Cope went to Washington to lobby for him, but the establishment (with some exceptions, including Joseph Leidy and the Harvard botanist Asa Gray) came down hard on the politically vulnerable naturalist. Professor John S. Newberry, a member of Marsh's advisory committee, characterized Hayden to Representative James Garfield of Ohio as "so much of a fraud that he has lost the sympathy and respect of the scientific men of the country." Newberry allowed that his former associate had "caused much good work to be done by others" but concluded that he was now "simply the political manager of his expedition." President Charles Eliot of Yale was less circumspect. "I have often heard his ignorance, his scientific incapacity, and his low habits when in camp, commented on with aversion and mortification," Eliot wrote to none other than President Hayes. "He has never shown that he is himself either a geologist, a topographer, a botanist, or a zoologist." Powell joined in, accusing Hayden, in a letter to Garfield, of wasting "splendid appropriations upon work which was intended purely for noise and show—in photographs, in utterly irrelevant zoological units which were designed to be scattered broadcast over the United States and Europe with his name blazoned on them."

Powell's lobbying for the new survey meant that he could not be considered a candidate for its directorship, but another job awaited him. An obscure provision of the survey consolidation act established the associated Bureau of Ethnology, and he became its head.

He and Marsh threw their support behind Clarence King, as admired by Eastern insiders as Hayden was despised. King "had everything to interest and delight" them, according to the consummate insider Henry Adams. "He knew America, especially west of the hundredth meridian, better than anyone; he knew the professor by heart, and he knew the Congressman better than he did the professor." In April, Secretary Schurz appointed King director of the U.S. Geological Survey, fobbing off Hayden with a sinecure that let him complete his reports but gave him nothing else to do. The specialist scientist had replaced the generalist naturalist in government as well as in academe. Hayden retreated to Philadelphia, where he developed *locomotor ataxia,* a nerve paralysis probably caused by syphilis contracted during Western expeditions. He died in 1887 at the age of fifty-nine.

King had no intention of running the survey for long, however. The $6000 salary was hardly enough to support him in the *beau monde,* and he expected to make a fortune in Western mining. In late 1880, he turned everything over to Powell and departed, which was just as well for the survey, since he had accomplished little. Powell would accomplish a great deal over the next decade. He "understood the congressman, as it turned out, better than King . . . ," Stegner observed, and "had no ambition to get rich. If he had any single ambition, it was the remarkable one of being of service to science, and through science to mankind." The major directed his survey toward three practical goals: determining U.S. geological resources, providing information for the expansion of the Western railroad system, and developing a system of irrigated agriculture. He did not entirely jettison natural history; at least not paleontology. Knowing that Marsh's scientific and academic connections would be vital, Powell immediately set out to appoint him the survey's vertebrate paleontologist.

Marsh was reluctant at first to take the position, because the survey law gave the government control over fossils collected by its expeditions. Once Powell assured him that no one would take his fossils away, the Professor joined up in August 1882. He would not

John Wesley Powell

have been O. C. Marsh if he hadn't, since it turned out to be almost as profitable as his Peabody-Yale deal. During the next ten years, the government would pay around $150,000 for a survey paleontological staff of fifty-four full- and part-time field collectors and lab workers, as well as a $4000 annual salary for Marsh, which he put back into survey work. It also would pay for lab expenses and for the illustration and printing of the three massive monographs that Marsh published between 1885 and 1896. Marsh not only had reached his goal of being the nation's top paleontologist; he'd become the best-financed vertebrate paleontologist of his time—perhaps of all time.

With so much government largesse added to his academic advantages, Othniel could use his private fortune to get every bone he remotely desired—and to live like a prince. In 1876, he had begun an eighteen-room, three-story brownstone mansion on a hill above New Haven, using $30,000 the Peabody estate had earmarked for a home. He decorated the interior so richly that it wasn't finished until 1881. "Its most striking room," wrote Schuchert, "was the high octagonal reception hall which he called his 'Wigwam'; he relates that when Red Cloud came into it, he looked up as if in search of the hole where the smoke went out." Marsh evidently designed the room as a setting for his anecdotes. "To the left of the entrance," Schuchert recalled, "was a very large round oak table of special design covered with western memorabilia from which Marsh loved to pick up the peace pipe that he and Red Cloud had smoked in 1874, the Mormon Bible that Brigham Young had given him, or some other interesting souvenir, and talk about it to his guests." The "wigwam" had a bewildering quantity of art objects—paintings, Japanese and Chinese enamelware and bronzes and *kakemonos*—and all the other rooms "showed the same astonishing abundance of bric-à-brac." Marsh later added a conservatory for a rare orchid collection, which cost $3600.

Marsh "liked extremely well to be with the rich and the great," Schuchert continued, "and he often was." His guests were a cross-section of eminent Victorians, including the co-discoverer of natural selection Alfred Russel Wallace, the polar explorer Fridtjof Nansen, and the inventor Alexander Graham Bell. The house was the setting for lavish entertainments, including a reception for the entire National Academy of Sciences when Marsh was sitting president in 1883. The Professor also held sway at a number of private clubs, "where he was regarded as a *raconteur* of parts," according to Schuchert. The most exclusive was the fourteen-member Round Table, which dined once a month at Delmonico's, in New York. Clarence King was one of the fourteen.

Marsh continued to be a celebrity, although a less sensational one, now that scientific administration absorbed him, which held limited interest even for the highbrow press. The *Times* and *Tribune* reported on such things routinely but briefly, and the sensationalist papers did so rarely. Sensationalism was moving on from the Wild West. The self-exiled Commodore Bennett now concentrated on spectacular international stories, such as two arctic expeditions that the *Herald* sponsored from 1878 to 1882. One of those, an attempt to reach the North Pole by the *Jeannette*, a ship named after Bennett's sister, became a notoriously gruesome disaster. Most of its members, including the *Herald*'s reporter, died of starvation. (Rival papers alleged that Bennett's rescue party searched the dead correspondent's pockets for unpublished dispatches after exhuming him.) Professor Marsh's prairie exploits must have seemed small change in retrospect. For his part, Marsh was content to be out of the sensationalist spotlight, now that it had served its purpose, and he was leery of public scrutiny. "Newspaper men were particularly guarded against," wrote one of his Peabody lab assistants, "even the editors of the college papers."

As Marsh reveled in opulent celebrity, Cope wallowed in frustrated obscurity. Like a grizzly shouldering a younger bear away from a salmon riffle, Marsh had monopolized government paleontology and was fiercely territorial. "Not only was he eminent, powerful, and incorruptible," Stegner wrote, "he was also, in his political and scientific rivalries, peculiarly mean, intemperate, and vindictive." Cope had about as much chance of a position with the new U.S. Geological Survey as did Red Cloud. His chances of support from universities and professional organizations were not much better, because Marsh influenced or controlled most of them. The new survey paid for the 1883 publication of a thousand page monograph on 350 species of Tertiary mammals, nicknamed "Cope's Bible." Edward had begun it under Hayden, but government support for his work ended there, though he kept trying to get Powell to publish a second volume.

Thrown back on his income for professional as well as personal expenses, Cope came up short. In 1878, he'd overreached himself by buying control of *The American Naturalist*, a move that backfired embarrassingly when thirteen scientists (including the venerable Asa Gray and James D. Dana as well as Marsh, Alexander Agassiz, and J. S. Newberry) issued a circular dissociating themselves from the journal. When that expense was added to the costs of preparing and illustrating the fossils he'd already collected, little remained for further collecting. "The summer of 1879 saw the last of Cope's purchases of important specimens from the field," wrote Osborn. In effect, Marsh's pre-empting the U.S. Geological Survey shut Cope down as a private fossil collector as well as a public one.

Edward made his final Hayden survey expedition that fall to the Pawnee Creek Miocene beds in Colorado, a trip "full of incidents characteristic of all his previous journeys," as Osborn elegiacally observed. He'd been to the site in 1873 and had "picked the ground clean" of fossils. "Mama will remember what great success, I had that season," he wrote to Julia. He was anxious about whether his past luck would continue, so he was pleased to find "a great abundance again"—twenty-five species in three days. Not as large a find as before, but some of the fossils were much better, including two saber-toothed tiger species. "This camp is on a stream that runs through the small bad lands here," he continued, "and is full of rushes, rose bushes, and willows. There is not a tree in sight, only the rolling plains covered with grass. Coming we saw at least 60 antelopes near here. Coyotes make the night musical and skunks make it fragrant. A large wolf prowled around last night."

Osborn wrote that Cope's letters showed no diminution of "his indomitable energy and enthusiasm about his discoveries . . . his ever prevailing humor and optimism." Yet a note of foreboding crept in. "My only objection to the region," Cope wrote, "is that it is part of the range of the Cheyennes and they are dangerous beasts. They have made two attacks on the borders since I was here, and have killed a good many people. I am now at the point nearest the Indian trail and tomorrow I go back a short distance to a safer camp, where I will stay a day or so."

Hostile Cheyennes didn't appear, but something nightmarish did. Cope acquired a field hand with qualities weirdly reminiscent of *Laelaps,* the dinosaur he'd described as the "devourer and destroyer" of all it could lay claws on. It was as though he'd been appointed as Edward's nemesis. He told Julia that he'd had to leave his gold watch as security to hire "a team and man . . . and now that the man is out here, he says he is not getting enough! . . . My indignation has been kindled against him for some hours." Three days later, he complained: "I had a disgusting time with my man. My lovely cook always remained in bed till I made a fire and cooked his breakfast! and then he rewarded me by eating most enormously. In four days he ate nearly all of the five large loaves of bread and was entirely above eating bacon. He ate nearly all of the only cans of peaches and condensed milk that I had! . . . When the bread gave out he drove me twenty miles against my orders, into a cow herder's ranch to get more!"

Cope's complaints read a bit like silent film subtitles, and one can imagine a professorial Buster Keaton helplessly waving his arms as Fatty Arbuckle drives him across the plains. The human *Laelaps* compounded the injury by being a "very poor driver" and worse guide. "Going to the ranch he had to depend on me for directions, for the distance was too great and we had to camp in a desolate place where the cold wind swooped down on us at night." It was so cold that the water in their canteens froze solid, "and we had to cut our bread with a hatchet!" When they finally reached the ranch, Edward raged, "I had to bake the bread myself while the [illegible] cook sat in the sun." He solaced himself by reflecting that he "did not suffer so much with the cold as the great cook who with barbarous providence has not enough clothes," and concluded that "the expedition was a success and perhaps the new lesson in human badness is worth something."

12

The Slippery Slope

With his genteel background and sanguine temperament, Cope probably didn't realize in 1880 just how badly he'd been beaten. If he had, he might have been more conciliatory and saved himself a lot of trouble. He probably could have sidestepped the Marsh steamroller by exercising some patience. His capital remained intact; his inventory showed he had about a third of a million dollars in stocks and bonds. He could have scraped along on his income until he obtained other paying positions. Although he'd been cast out of the insider establishment, he retained a substantial scientific reputation and many friends, and he already had more than enough fossils to study for the rest of his life. He might have eventually ingratiated himself with Powell or at least with the Smithsonian's Spencer Baird, a less adamant Marsh ally.

Conciliation was not the way of a man who'd never had to apply for a job he really needed, however. How could Dr. Edward Drinker Cope pinch pennies and curry favor as the upstart Marsh bestrode the field? *He* had to have power, too, and power was money. Cope decided to invest his capital in Western mining ventures and thus make a fortune. He had experience in metal-prospecting going back to his 1860s Appalachian expeditions, a good knowledge of geology, and a firsthand familiarity with the West. Other scientists, notably Alexander Agassiz, had made mining fortunes. Even the riffraff he hired as bone sharps sometimes struck it rich while prospecting. How could a man of his innate superiority fail?

Cope might have found a cautionary answer to the question had he been able to foresee the career of the great Clarence King, who went into Western mining with much more justified confidence. King's specialty was practical geology, and he had access to major sources of capital in the East. "Whatever prizes he wanted lay ready for him," wrote Henry Adams. "With ordinary luck he would die at eighty the richest and most many-sided genius of his day." As it happened, King died penniless in an Arizona hotel at fifty-nine, the same age as his vanquished adversary, Hayden. His Mexican mining ventures did make money, but King spent it lavishly, and it wasn't enough for him to survive the Gilded Age's stock-market plunge. An 1893 crash wiped him out, and he passed his last years struggling against tuberculosis and mental illness. Adams had considered King's education "ideal, and his personal fitness unrivaled," but concluded that "the result of twenty years' efforts proved that the theory of scientific education failed where most theory fails—for want of money."

It seems unlikely that even King's sad tale would have deterred Cope. King, after all, was an easygoing *bon vivant,* and the residually devout former Quaker might have considered his fate the wages of sin. When his friend Persifor Frazer, a geologist in Philadelphia, asked how he managed to do such an enormous amount of work, Edward sententiously replied that he did so by "eschewing liquor and tobacco and minimizing nervous storms." Lessons of personal inadequacy can be learned only from personal experience, and Cope had little notion of what he faced. A relict Enlightenment gentleman jumping into mid-Victorian finance, he was like one of his majestic but primitive creodonts entering an arena of wolves and lions.

Of course, the financial carnivores hid their claws at first, and the years right after Marsh's ascendancy probably didn't seem too different from the ones before. Edward published seventy-six papers in 1880, a lot even for him, although he doesn't seem to have done much else, which does suggest some discouragement. He regained his confidence the next year, however, and sent seven collecting parties into the field. It was the summer that David Baldwin found the San Juan Basin Paleocene fossils that confirmed his new epoch. Cope combined a trip to scout mines in southwest New Mexico with visits to Baldwin and a collector named Jacob Wortman, who was finding more early mammals in western Wyoming's Bighorn Basin.

In New Mexico, a flash flood washed stones and trees across the railroad, and the train was so packed with soldiers that Cope feared their muskets would attract lightning. The soldiers were there to deal with Apaches. "I think sooner or later they will give some trouble around here," Edward wrote to Annie from the Lake Valley mining district in August. "They have been doing mischief all round us and in fact have been seen near the mines. I leave however today . . . The retirement from the pressure of active work is good for me, and I am sleeping wonderfully." He was enthusiastic about the Lake Valley prospects. "This is a great mining district," he continued, "perhaps the best in the entire west, and the miners are becoming numerous very rapidly."

Cope was right about the district. "Huge amounts of gold and silver were taken out," wrote Url Lanham. "In one mine, opened about the time Cope arrived, an ore vein led into a cave filled with $3 million worth of pure silver chloride." But wealth and ease were not to be had immediately, and parts of Edward's 1881 trip must have reminded him of his Fort Bridger flounderings. On an excursion he took to search for some Wyoming Eocene fossils, the train stranded him at a desolate whistle stop, and he had to sleep on the platform. "A cold wind sprang up and gave me the creeps," he told Annie. "I was wakened once by an animal fussing around my head. I found it was a friendly cat, and I was glad to take it in for the warmth."

The next summer was even busier, as Cope led a party of investors to Lake Valley. "The Philadelphians are the best of the lot

—decidedly," he wrote. "We don't gamble, and but few play cards in any way, but the Californians are at it constantly." He pronounced the party well pleased with the mines, in several of which he'd already invested. Cope by now was such an important figure that he was asked to address the local army detachment. "I do not approve of war but in this country the soldiers are the only armed police, and they have to be," he reported. "I made a speech and they cheered me." Then it was on to El Paso to check mining prospects in Chihuahua, "which I would have gone to see had I not been afraid to go too near the Apaches," Edward reassured Annie. He added that he'd shaved off his beard "except moustache and goatee as an experiment, and I think I look like a gambler, or some other piratical craft."

Pursuing his 1882 mining investigations to Idaho, Cope crossed the desert in a stagecoach that ran past the still undiscovered Pliocene horse fossils of the Hagerman Fossil Beds. "Sage brush or nothing grows on lava or sand," he wrote to Annie. "We dined at 'Roast Hog' where they bring water 15 miles, and took supper at Arco where the mosquitoes were in millions and nearly ate the supper for us, but for a Chinaman who burnt a lot of old hats and made such a stench that the mosquitoes got out of the dining room." Idaho mining didn't pan out, though, nor did a paleontological foray to eastern Oregon. Cope had heard of bones at Warner Lake, and got a good outfit from the local army unit, but found no bones. Moving on to a sandy desert near Summer Lake, he searched in vain for fossils exposed by the wind. "The weather was excessively hot," he wrote, "and the cattle gnats got into my hair and whiskers and poisoned me sore, so that the surface was like the Rocky Mountains under the hair."

Cope spent most of 1883 preparing for publication his "Bible," the monograph on Tertiary vertebrates, and squabbling with John Wesley Powell. Wallace Stegner thought he "dragged his feet" on the book to spite Powell and deliberately added material so that more volumes would be needed. A letter to his daughter gives a different impression. "I have been here for several weeks looking up fossils and printing," Edward wrote from Washington in February. "The big book I have been working on for ten years is to be printed at once, and I am glad of it." Yet he evidently did suspect Powell of rushing to end his obligations to Hayden, and in May the public printer stopped work on the grounds that he did not have a finished manuscript. Powell and Hayden agreed to publish the material in hand, but the author objected strenuously, and the deadlock lasted for several months, until Cope ungraciously sent the remaining manuscript to the editor with a memo saying, "Can't we scotch Powell?" The editor, who was the geological artist William Henry Holmes, had been a Hayden man but "was not interested in puddling old blood," as Stegner put it. "He showed the letter to Powell, who could afford to ignore it. Cope was blocked in the government bureaus, and could do no harm."

In September, Cope returned to New Mexico, where his Lake Valley investments were languishing; the financial carnivores were starting to feed. "With childish confidence," recalled his friend Persifor Frazer, "he accepted statements of glowing prospectuses and interested promoters, and scattered his capital in many directions on insecure guarantees where he had neither the time nor the ability to guard it. In consequence with this he was almost immediately confronted with losses which alarmed and confused him. Instead of accepting these as severe warnings . . . and withdrawing on the best terms available, with a diminished but still handsome fortune, he plunged deeper into these investments and lost so heavily that his whole subsequent life was harassed."

Cope continued to sound optimistic in his letters home, telling his wife he'd been elected president of the mining company and had replaced a crooked engineer named Bunsen with a friend, the former Hayden survey geologist Frederic Endlich. "When I came here I expected to find things at a pretty low pass," he wrote. "I am agreeably and greatly disappointed. We will not go to the poorhouse yet, though I know some people that ought to go to jail . . . I feel that the trouble of the last two years is over and that I shall succeed in this enterprise." Those two years had already taken their toll. "I had too much on hand last winter," he wrote a week later, "especially of annoyances produced by others, and was nearly worn out when I left home."

The annoyances weren't over. After moving on to look at mining possibilities in Monterey, Mexico, Cope learned that the Lake Valley company had dropped him from his newly bestowed presidency. "I thought it rather rough under the circumstances," he wrote to Annie. Cope usually played down bad news, but an attack of fever in Monterey may have been a manifestation of his real feelings. His temperature reached 105 and his pulse rose to 110 before he recovered "by a judicious use of aconite, morphia, nitre, and quinine." Cope's fevers doubtless were malarial; the mosquito-vectored disease was endemic in much of the United States, and was likely to afflict anyone who camped out. But his case clearly had a psychosomatic side, his temperature rising in times of stress and contradicting the impression he tried to give of a man in control of his affairs.

Cope couldn't pay any field collectors that summer, though he managed a few forays on his own. "The only difficulty I can see is the expense, which I have reduced to the lowest terms they will stand," he wrote to Annie from the Cannonball River in Montana, "where badlands are said to be especially bad." He found only one dinosaur tooth in the Cretaceous beds he had come to explore, but on the bluff tops he chanced on a much younger fauna of rhinos, a "hog with legs to his chin," a small dog, a saber-tooth, a three-toed horse, and sixteen other species he thought contemporaneous with the White River fauna. He also found some of the last nonfossil megafauna on the Great Plains. "I lived on bison meat all the time I was out, except when I ate venison," he wrote. "There are several hunting parties out and the game must soon be reduced in that country."

Cope then looked for bones in New Mexico, where he found nothing much except for a rhino skull he bought from another collector. He didn't seem to mind. "This kind of work suits me amazingly," he wrote, "and I only wish mining matters had permitted me to have remained longer." He was happy just to be in the field. "I am writing by a fire of dry cedar sticks which give a good light," said an October letter from New Mexico's Mogollon Mountains. "The moon lights up the wall of the canyon, and the rush of the water never slackens its noise. Besides this, nothing is audible save the crickets and the noise of the horses' feet as they occasionally change position."

Some of Cope's field pursuits may have been less innocent. He acquired a reputation as a womanizer. A letter written by Henry Fairfield Osborn in 1930 to a Princeton University Press editor mentioned "many tales of ladies," and quoted the artist Charles Knight as saying that Cope's mind had been "the most animal" and his tongue "the filthiest" he'd encountered, and that "no woman was safe within five miles of him." In fact, Edward's letters to his wife often contained admiring comments about women seen or met during trips, comments that increased in inverse proportion to the husbandly endearments typical of his 1870s letters. Boorish as it was, this epistolary ogling may have been part of fantasy rather than behavior. His biographer Jane Pierce Davidson found no real evidence of philandering. Still, nearly all of Annie Cope's letters to her husband have disappeared, hinting at marital problems, and the couple spent less and less time together in the 1880s.

When Cope returned to Lake Valley early in 1884, he was evidently in an ugly mood. A March 29 letter from Kansas complained of Jews who traveled "with huge families in the sleeping cars. I have had them all the way out. They are often unmannerly people. The men go about half dressed and smoke everywhere." Since he was thinking of selling his fossil collection, he stopped at the University of Missouri to discuss it. "I am very sorry to put it in such an out-of-the-way place," he wrote, "but if they have more enterprise there than anywhere else they will get it." Selling the fossils he'd amassed with such enthusiasm was a measure of desperation, evident in an April 10 letter from Lake Valley. "With mines it is either everything or nothing," he told Annie. "I went in hoping to increase my income. I did so for a while; then it fell to nothing."

Being at the mines cheered him up again. Several were doing well, as was a mill he'd helped to finance. "Altogether it looks as though we should begin to rise from the depression in which we have fallen," the April letter continued. "But this is going to take time. The development work which we could have done once with ease, now takes time and labor, for the old board foolishly spent its money." Considering that the Lake Valley situation always seemed to improve when Cope was there and deteriorate when he wasn't, he might have wanted to stay for the summer. A letter to his daughter outlined a pleasant daily routine of inspecting the work and drilling for new ore bodies, interspersed with evening conversation and a little fossil-collecting. Yet Cope spent less than two weeks at Lake Valley before hurrying off to Mexico City, where he promptly succumbed to fever again. The attack was milder than the 1883 one, but he couldn't eat for three days.

The next year Cope did the same thing, except that he lacked the energy for fossil-hunting. "I have the prospect of short money for some time to come," he wrote to Annie from New Mexico. "I have not yet gone fully over the mines here, but I have seen enough to know that we are not making anything but our expenses." Again, he tried to be optimistic: "We have before us an excellent project which will put us on our feet perfectly if it goes through." Then the letter's forced hopefulness faded to weary confusion: "I fell among . . . thieves here, but by great exertions defeated them. Now I am among imbeciles." Two weeks later, he was in Mexico City. "I have so far escaped malaria, thanks to the medicine I brought, I suppose . . . ," he wrote. "But I do not wish to remain long. I have no appetite, and little energy of body." He wandered around, dazedly admiring the tropical forest in Veracruz. "My health is very good and all tendency to fever is absent," he reported ten days later. "In this respect my trip has done me a world of good, and not less the relief from the horrible financial anxiety I was in. In a few weeks, I expect the latter will take some definite shape which it has not now."

The downhill slide of the previous five years continued, however. In May 1885, he wrote to Osborn, recently named professor of geology at Princeton, asking whether he wanted to hire Sternberg and another collector to work on a Kansas fossil bed: "I cannot employ these people, so I would rather see you and Professor Scott have the chance than anyone else." Sternberg ended up working for Marsh. Cope didn't do any collecting that summer, and money got so low that he had to put his 2100 Pine Street residence up for rent and move into small rooms nearby. The anxiety drove him to still another flight, to Yellowstone with an excursion arranged by the American Association for the Advancement of Science. "Absence from home at this time is not very favorable for renting the house," he wrote to Annie from Wyoming, "but I considered the benefit to my health would be great enough to justify my taking the trip. I hope however that between thee and Mr. Ritter it may be rented in my absence for $1000—for the next year. *If properly packed,* the greater part of the furniture should go into the small third-story room and the kitchen of 2102 Pine Street."

Cope's mining career was over, but it took him until December 1886 to face the fact. "I found that certain things had occurred in connection with the company which will put them into tedious litigation for a long time which will use up any profits that will be made out of the ore," he wrote to Annie after talking to one of the trustees in the "S. Grande" Company in New York. "I concluded that the stock would probably become worthless or nearly so for a time at least. However, I had not the money to pay the assessment." He went to his stockbroker, who "offered to sell it for me for more than the trustees were giving for it. That was the equivalent to his buying it from me at that price. I gave him the order to sell." The financial carnivores he'd hoped to emulate finally had picked Edward clean.

"In any case," his letter continued numbly, "I must have something to live on this winter, or until I can get something from the Govt. or the Tilden Estate. All I can say is that I thought best to sell my large lot, as long as I can get anything for it. Please *do not tell* that I am trying to sell, as it might injure the sale, or prevent it."

With his capital gone, the only arenas in which Cope could hope to retain some of the high status he'd always taken for granted were those from which he'd haughtily withdrawn: academe and government-sponsored science. Of course, paleontology jobs were scarce, especially for fallen grandees with turbulent reputations. The vague possibility of the "Tilden Estate"—that he might head some kind of philosophical institute—came to nothing. Princeton had rejected his application for a teaching post on the grounds that his opinions were too radical. In government, aside from crumbs like work for the Canadian Geological Survey, the only substantial possibility was doing another volume of the "Bible," which John Wesley Powell reluctantly had published in 1884. This would not have been a good prospect. Powell maintained that Cope's tome had far exceeded its intended scope, and that it had exhausted the entire appropriation for Hayden Survey publications. Yet Edward was nothing if not persistent when he wanted something badly, and he pursued his "New Testament" doggedly through the late 1880s.

If Powell wouldn't furnish the money, he'd go to his boss at the Interior Department. "I had my interview with Secretary Lamar at last," Edward told Annie in May 1886. "He was pleasant and asked for full written explanations, and that I should write a letter *for him* to Maj. Powell. This I have done and have sent the entire matter in to him. It remains to be seen what will come of it." Nothing came of it. Two weeks later Cope was back in the Interior Department waiting room with the other petitioners and job-seekers. "I went 5th day to the office of the Sec. Interior and waited two hours without seeing him," he told Annie. "Next day I waited one hour, and then Gen. Warner came along and saw the Secty. himself on my account. He made an engagement for me to go tomorrow, and tomorrow I go with considerable hopes of seeing him."

It was not to be. "I spent 5½ hours today waiting for the Sect. to receive me," he wrote the next day, "and although he appointed today he failed to do so. I go tomorrow again, and so on! I was not idle all the time; I read some proofs and had a book to read also; nevertheless I am a good deal disgusted." Cope didn't get to see Lamar again until June 9, and the only result was that Lamar dictated a letter to Spencer Baird at the Smithsonian "asking information and opinion about my work. The Sec. looked as if he were at a funeral as he dictated the letter."

If Lamar wouldn't appropriate the money, Cope could go to Congress, but even he could see that without active support from the administration, there was little prospect of a bill to finance another book on Tertiary vertebrates. "I suspect that my chances are over for this session of Congress," he wrote. "The conclusion is I return here next winter and try it over again, and commence *earlier*. . . I am not happy at the result of my experience, but I am glad to be rid of the suspense and exertion for the present." Cope evidently couldn't get enough of waiting rooms, however. He kept lobbying for most of the summer, and convinced himself he was near success in getting funding for the book through a committee amendment to an appropriations bill. "So far as words go, my appropriation is secured!" he wrote to Annie in late June. "But I do not allow myself to be sanguine about it, or to expect anything until I actually see it done."

It was just as well that Cope didn't expect anything, because after another grueling month, the whole thing dissolved like the Potomac morning mist. "*My bill is thrown out* on a technicality, no doubt in the interests of Powell," he wrote incredulously. "This means a great deal to us. First, I will have to be here again next winter. Second, we will have to live very cheaply for a year." Since the Copes already had been living very cheaply for more than a year, Annie's feelings can be imagined. "Of course I am much disappointed," the letter continued. "Nevertheless I am pretty sure to make it next winter, 'they say,' and I have certainly met with very little direct opposition from Congressmen." Given Cope's political status, that was like a flea saying it had met very little direct opposition from an elephant.

Edward was back in Washington that fall, and, in February 1887, actually managed to get his bill through the Senate, but that was as far as it went. "I received as many compliments as I could expect for my work from the country as well as Europe," he wrote, "but that is my greatest sin in the eyes of certain people. This class are jealous, and 'what is more cruel than envy'. . . As for my position here, Professor Baird is afraid of me, I suspect, and will give me nothing regular yet awhile." But the next fall he was hopeful again. "I *may* get it!" he wrote to Julia in December, "and then paleontology will begin again." He didn't get it, but in 1888 he found hope in the appointment of a new secretary of the interior, although he told Julia in February that he already had called on him three times without seeing him.

It took Cope longer to see the new secretary, Vilas, than it had to see the old one, and the results were even less encouraging. "Today I had an interview with the Secretary of the Interior," he wrote in March, "and received the cheerful assurance that he knew nothing of the subject, and had not had time to examine it. But that if Maj. Powell recommended it he would then give it his serious attention! As Maj. Powell is the author of the difficulty, I concluded that my chance of finishing the work of the Hayden survey is very slim!"

Cope *still* didn't give up. He went to see Powell as Vilas had advised. The wily major took him to lunch and agreed to present the matter to Vilas, but evidently didn't do so enthusiastically. "Powell says he cannot get Vilas to look into it," Cope wrote to Julia a week later, "and he says he will not try any more! But he proposes to me to publish in his Survey on condition that I give a set of fossils to the National Museum, and will take up the Permian fauna and the fishes in general, first. This will be a good beginning, and as he proposes to pay a salary, I feel much relieved." Powell was dangling straws, however, and the fact that they "relieved" Cope was a measure of the latter's desperation. His collection was his only substantial remaining asset. "My entire future in a financial sense, which means thine and Julia's so long as she is single, depends on that collection so far as I can see now," he'd written to Annie in 1887. To give away a good part of it in exchange for a minor, temporary position under his archrival's ally would have been madness.

Nothing came of Powell's self-serving offer anyway. Cope's "New Testament" also had reached a dead end. The year 1888 brought to a crescendo the drumbeat of rejection and exclusion that had hammered him through the decade. "I had a letter from the people of the Central Park Museum declining my services and my proposition to sell my collection," he told Annie in June. "As this is the second time they have done that, I think we can consider that place is finally out of the question." A few days later, he wrote that another possible buyer had turned down part of the collection, and that even plaster casts of his fossils weren't selling, "except one to be paid for, *perhaps,* after July 1." It also appeared that he might lose ownership of *The Naturalist.* He'd been forced to get a new publisher the year before, and that had brought only "endless trouble."

Cope was living in Chaplinesque poverty by this time. "I had many queer experiences in the odd, half-bohemian restaurants which the naturalist frequented," wrote the patrician Osborn, heir to a railroad fortune much greater than Edward's had been. "At the Philadelphia market near Pine Street we could always secure a good meal at twenty-five cents, including a large slice of roast beef followed by mince pie. Then Cope would bring out a venerable cigar case from which he would extract a half-burned cigar. This was smoked following our coffee, then replaced in the cigar case for future reference. Small economies of this kind were shown in the case of paper, and twine, in fact everything about the study."

Osborn stressed Cope's good humor in the face of adversity, but poverty can hardly be laughed off. In April 1888, Edward couldn't afford to accept an invitation to bring Annie and Julia to the annual convention of the National Association for the Advancement of Science in Washington. "Several persons asked after thee," he told Annie. "I didn't tell them the reason why thee and Julia were not there. If I had the $ to have had you come, you would have been there certainly. But I will be short, I fear, til July. Then I hope for a change for the better."

There was no change by July. "The financial situation has not improved," Cope wrote to Julia in September, "and I will have to do all possible things to get through this winter, when I hope better times will arrive." Nor was there a change by the spring of 1889. "I am just trying to raise the money to pay $500—interest on mortgages on 2100 and 2102 Pine St. due in June and I have a like amt. to raise in October," he told Julia. "I do not know where it is to come from, but I shall have to do some begging and then get only part of it, I suppose." Cope had learned what most poor people grow up knowing—that things will probably never improve. "The ethics of business are peculiar and the disposition to get away from others is the usual spirit of it," he'd written to Julia the previous year. "For it is true that well-to-do people can generally get more pecuniary aid than those who are in a tight place!"

13

Behind the Arras

If Cope hated Marsh in 1880, his feelings toward his affluent enemy as the decade continued can be imagined. The two men doubtless would have kept squabbling even if Cope hadn't failed so painfully, but such animosities often trail off into mutual weariness if each antagonist has some territory. The game changed when Marsh took everything. It was no longer simply a conflict of personalities; it was victor and victim—something evoking more primal energies—and it passed beyond competition, which can remain rational, and entered the realm of vengeance.

Like many voluble people, Edward used language to hide his emotions as much as to reveal them. "Few men have succeeded as well," wrote his friend Persifor Frazer, "in concealing from any one friend or relative, however close, all sides of his multiform character." His written remarks about his enemy were guarded. "Inferior men like Powell and Marsh may have great influence simply because they have gotten position," he wrote his wife dismissively. "It makes little difference how it was done." He cultivated the image of a happy warrior driven to oppose the Marsh cartel by a love of truth, justice, and a good fight rather than personal hatred.

Edward did like a fight. Henry Fairfield Osborn recalled that Cope and Persifor Frazer once got in such a violent argument in the Philosophical Society that they went outside and came to blows. "On the following morning I happened to meet Cope and could not help remarking on a blackened eye. 'Osborn,' he said, 'don't look at my eye. If you think my eye is black, you ought to see Frazer this morning!' " Osborn thought his friend had a "humorous Celtic attitude" toward the Marsh conflict, and enjoyed it for its own sake. Cope "particularly relished all new anecdotes and jokes relating to his distinguished rival," he wrote, "but as a matter of fact showed far more humor than bitterness." According to the paleontologist E. C. Case, whose father had been a contemporary of Cope's, he "met honest opposition with a vigor honoring his foe, but fraternized cordially after a battle."

But Case also knew of the happy warrior's darker side. "He hated opponents he could not respect," he wrote, "and gave them heavy blows how and when he could." In two Cope anecdotes, Osborn half-consciously corroborated an impression of calculated vindictiveness. "One day," he wrote, "he slyly opened the lower-right-hand drawer of his study table and said to me: 'Osborn, here is my accumulated store of Marshiana. In these papers I have a full record of Marsh's errors from the very beginning, which at some future time I may be tempted to publish.' " Another time, one of Cope's scientific names puzzled Osborn. "I had diligently searched the Greek dictionary for the term *cophater*," he recalled, "because I had always found Cope's specific, as well as generic, terms highly consistent. He remarked: 'Osborn, it's no use looking up the Greek derivation of *cophater*, because it is not classic in origin. It is derived from the union of two English words, Cope and hater, for I have named it in honor of the number of Cope-haters that surround me.' "

Such caprices are reminiscent of a Webster or Shakespeare character—and not one of the pleasanter ones. Wallace Stegner, a Powell partisan, grew eloquent on Cope's balefulness. "The geological survey was very truly a consolidation, and contained men of all four of the original western surveys in its personnel," he wrote, "but one man it could not placate was Professor E. D. Cope. He took over Hayden's place as the leader of anti-Powell forces among scientists . . . Cope was a character out of fiction, a distinguished scientist with an emotional life like that of a villain in Jacobean tragedy. The very bones of Tertiary mammals, as he cleaned and arranged them in his Philadelphia home, cried out to him 'revenge!' Vanity and hatred stained Marsh's career, but they utterly corroded Cope's."

Stegner demonized Cope, who, despite many flaws, was not an utterly corroded villain but a noble man in his way, as were Powell and Marsh. Cope would have done less damage if his main qualities had been vanity and hatred—villains are always clawing in corners; it is when nobles start doing it that the roof comes down. Yet there's something to be said for Stegner's demon Cope. As the 1880s ground along, Edward became noticeably more like the stealthy bogeyman with whom Marsh had frightened his Como Bluff collectors. This was an extension of something that always had characterized their relationship. Just as Cope learned from Marsh in the 1870s to be a capitalized, fossil-grabbing scientist, he learned in the 1880s to plan devious political strategies. He became Professor Moriarty to Othniel's academic Sherlock Holmes.

Cope might not have been a threat had Marsh been as firmly established as he seemed, but the Yale Professor's eminence and gala social life had a strange one-sidedness. Henry Farnam, Marsh's colleague and former student collector, recalled that Othniel had kept his telephone "unplugged all the time when he wasn't using it because he'd had it 'put in for his convenience and not for that of others.'" Despite his fame as a *raconteur,* some clubmen thought him a bore, calling him "the great Dismal Swamp." His bachelorhood troubled society. A class reunion speaker joked that Marsh stayed single because he couldn't be happy with less than "a collection of wives," but his only known *amour* after 1870 was "the daughter of a Baltimore railroad magnate," which ended, Schuchert wrote, "because of the lady's displeasure at a too marked display of devotion on the professor's part." Marsh enigmatically expressed regret at this by telling his Peabody assistant John Bell Hatcher that his house "needed a 'garret' to make it complete."

Homosexuality has been suggested as a reason for Marsh's secretive bachelorhood, and Othniel certainly succeeded in surrounding himself with young males. Jane Pierce Davidson speculated that Cope had been referring to sex in 1873, when he wrote to his father that Marsh was "not normally or properly constituted." There is no evidence of any sexual activity in Marsh's life, however, or of much emotional contact of any kind. "My belief is that he always lived apart from those about him," wrote Timothy Dwight, the president of Yale. "I doubt whether even his most intimate friends penetrated the recesses, or really in any measure understood him . . . I question, indeed, whether he had intimate friends."

In one social area, Marsh had developed a dangerous weakness as a result of his 1880 triumph. The huge volume of work connected with Powell's survey required that Marsh hire more assistants, and since he was as careless as ever about supervising and paying them, and as domineering about getting all the scientific credit, his lab soon burst with unhappy helpers. His way of working with assistants also contributed to the unrest. He would assign each one a series of fossils and tell him to study their evolutionary affinities. "In due time," wrote Charles Schuchert, "he would sit down with each man, asking questions, discussing conclusions arrived at, and all the while making notes in regard to what he heard and what he saw in the assembled material." Information flowed in only one direction, however. Marsh kept "his conclusions strictly to himself in these conferences, especially when he differed from his staff; he rarely exposed his final conclusions until they appeared in print." Obviously, this gave rise to suspicions as to how he had reached his published conclusions.

Marsh had alienated the originally hero-worshiping Samuel Williston in little more than a year. "I had begged permission to study the fossil fishes of Kansas," Williston wrote, "with the expectation and hope of doing some research work for myself, knowing that Marsh had no desire nor intention of ever doing any work on them himself. He assented and set me to work studying fish skeletons, done on my own time, of course. After some months' work upon the recent fishes I suggested that he let me take some of the specimens I had collected to my room to work upon evenings and holidays. But he declined, giving me to understand that he did not want me to do any paleontological work for myself . . . I recognized then, or a little later, that I would never have the opportunity of doing any independent research work in paleontology as long as I was his assistant."

Another Yale graduate assistant, Erwin H. Barbour, also conceived an active hatred for the Professor. "Not only does he avoid helping his assistants to better positions in geological fields," Barbour would write in *The American Naturalist,* "but he often hinders them by trampling on their good names when gone. We assistants watched the evolution of a falsehood from his lips, from the day when he said 'that man has resigned,' to the month when he said 'I had to let him go; he was a bad lot,' until still later he 'dismissed him because he was unreliable and light-fingered.' "

Williston and Barbour used their Yale positions to get higher degrees and move on, but Marsh completely subordinated the career of the less aggressive Oscar Harger, his assistant since the student expedition days. A gentle, dwarfish man with an enlarged heart, Harger was married and lived on his salary, so he never was able to do independent work except to write some papers on invertebrates. Harger's plight became a *cause célèbre* among the other assistants and Yale faculty, who vainly urged Marsh to let him be named co-author. This unrest ran deeper than pique at Marsh's autocratic attitude. Some observers thought Harger did much more than fetch and carry fossils, the unobtrusive role he'd played during Huxley's 1876 visit.

"Harger was Marsh's eyes and brain," wrote Url Lanham. "Marsh, after his student days, read little in vertebrate paleontology. Harger read everything. Harger could relate the facts discovered in the laboratory to the contemporary world of paleontological research, and he provided sound judgment that helped Marsh convert the work of the laboratory into the standard scientific commodity, publication." The situation led to speculation as to who really had written some of the publications, and after Harger's 1887 death from heart failure, Williston did more than speculate. "To my personal knowledge," he wrote in an *American Naturalist* obituary of Harger, "nearly or quite all the descriptive portion of Professor Marsh's work on the Dinocerata was written by him and was published without change, save verbal ones. The descriptive portion of the Odontornithes was likewise his work, but this I cannot say from personal knowledge."

With eighteen sulky employees at work unpacking, cleaning, repairing, and cataloguing bones under the Professor's saturnine eye, the Peabody lab seems a macabre parody of Santa's workshop. A photo of Barbour and three other assistants is like an Edward Gorey drawing—grim young savants in side whiskers, frock coats, and wing collars drawn up around a grinning brontothere skull. Marsh did little to dispel the venomous atmosphere. Schuchert observed that his "drive and determination spent themselves as his wealth increased and his contacts widened . . . He had become overwhelmed and confused by the very mass of his fossil riches and by the effort required to direct his superabundant staff in the laboratory and in the field. At times, his assistants were left for a day or more with nothing to do except talk over their grievances while Marsh lingered in New York at the University Club or the Century Club, where he undeniably liked to 'spread himself.'"

One change that Marsh did make only aggravated the unrest. He began using ambitious professionals as assistants instead of dependent students, hiring three German paleontologists, named Max Schlosser, Otto Meyer, and George Baur. According to Url Lanham, he did so because of his nostalgia for "the Prussian laboratories of his student days in Europe, where a single professor ruled as undisputed autocrat over assistants who loved to be subservient to capricious and arbitrary authority." But the measure backfired. Schlosser and Meyer soon left, and Baur proved an active rival instead of a Teutonic thrall; he published seventy-five papers on fossil and living vertebrate embryology while in Marsh's employ. Marsh actually helped Baur by lending him money and sending him to

Europe to study fossil reptiles, so his motives for hiring the Germans may have been idealistic or sentimental as well as calculated. Schuchert thought "that Marsh recognized Baur's great ability, and that he went much farther to keep him satisfied than he had ever done with any of his other assistants." Baur, however, looked down on the Professor's "mental equipment and knowledge" and coveted his fame, while his indebtedness poisoned their relations.

When they arrived, in 1884, the Germans acted as catalysts for the caustic reactions that had been brewing. Baur was touchy and arrogant, Otto Meyer was uncooperative, and Max Schlosser seems to have been a caricature of the swaggering Hun. He once threatened to cane a large group of undergraduates who jostled him as they left a building. When he returned to Germany, in the spring of 1885, the Peabody lab was in turmoil. Williston had written to Powell the previous fall to complain of Marsh's administration, and Powell had turned the letter over to Marsh, who scolded Williston for going over his head. "He was angry, perhaps properly so," wrote Williston, "and on his return from a trip to Washington accused me of things that were not honorable. In my anger I accused him of various things of the past not very creditable to him, that he thought I did not know, and suggested that it was about time for us to sever our relations. To this he agreed but asked me to remain till the first of July that there might be no outside talk."

There already was plenty of "outside talk." As the "Marshiana drawer" demonstrated, information from New Haven had a magical way of flowing to Philadelphia. If Marsh's New Haven lab was a sullen Santa's workshop, Cope's Philadelphia one was a vengeful sorcerer's den. A photo shows a solitary, goateed Edward glowering at a pair of skulls, a pose that may have inspired Stegner's notion that the very bones cried out to him "revenge!" There were plenty of bones; visitors described them as piled everywhere. Except for one assistant, a technician named Jacob Geismar, the lab's only other living occupants were a Gila monster in a tank and a tortoise that wandered around the floor. Intrigue flourished in this necromantic atmosphere.

Osborn and Scott, who visited the Peabody lab in pursuit of their own research, probably discussed their trips with their Philadelphia friend. Marsh certainly suspected them of doing so, and may have trailed the Princetonians in carpet slippers, peeking from behind bookcases and doorways, while his attendants showed them around. Cope himself stalked the lab, and this invasion of his *sanctum* was one of the things Marsh would denounce most bitterly during the *Herald* scandal. When Cope accused him of not allowing scientists to examine his collections, Marsh replied that the charge was "in part true. Visiting scientists of good moral character are always welcome, but I have learned caution by experience." One of the learning experiences, according to Othniel, had occurred during a Cope visit in 1882.

"It was a Saturday afternoon, when the workrooms of the Museum were closed and most of the attendants absent," Marsh told the *Herald*. "He was accompanied by Professor Silliman, then associated with him in mining enterprises, and who . . . had no right to enter any private room of the Museum. One door was accidentally open as an attendant was passing through, and Professor Cope and Silliman by that entered one of my private rooms, where the results of years of my labor were spread out ready for publication. Professor Cope began by uncovering the lithographic plates which he examined attentively, and then passed into an adjoining room and made a close inspection of many valuable fossils which were my own property and unpublished." Gad!

"Three more rooms were subsequently visited in the same surreptitious manner," Marsh continued. "These were my private workrooms, where some of my rarest fossils were stored. Many of these were covered, but even these did not escape him. One rare skull he exposed and carefully examined, the assistant in attendance not knowing who he was, or the consequence to Professor Cope personally would have been serious . . . This outrage was reported to me next day by the two attendants who had witnessed it, and I at once took prompt measures to redress, as far as possible, the wrongs I had suffered." Erwin Barbour wrote that after Cope's visit, "we were directed not to admit even Professor Silliman or any of the Yale faculty, much less a stranger—a demand so unjust that I for one refused, once for all, absolutely, to do anything of the sort."

There may have been truth in Marsh's allegations, though they may have been smoke to obscure embarrassments that Edward had dug up on Yale visits. Cope showed satisfaction in a bland reply to them. "By Professor Silliman's invitation, on one occasion, I walked through the Yale College Museum, including the rooms where Marsh's specimens were lying on a table," he told the *Herald*. "I saw little of interest to me except one specimen which was uncovered. I wrote nothing about what I saw until Marsh had published his final work on the materials then on the tables. Then I referred to the one specimen I had seen since Marsh had 'restored' it with plaster, in such a way as to deceive everybody as to its character. Marsh's excitement at this event amused everybody, including sundry professors at Yale. This will give an idea of what Professor Marsh regards as 'depredations' on his museum."

A chance for Cope to use such Marshiana materialized just as the Germans arrived. In July 1884, Congress initiated a commission to investigate the efficiency of scientific bureaus, including the U.S. Geological Survey. As the investigation crept through Washington during the next few years, Powell and Marsh didn't control it as well as they had the 1879 survey consolidation. Competition was growing with other bureaus, like the Coast Survey. Reactionary politicians like Alabama Representative Hilary Herbert, were growing wary of Powell's power, asking themselves, as Stegner put it, "On what meat does this our Caesar feed?" Cope was working to exploit this dissatisfaction by 1885. "I had a letter a few days since from Mr. Williston of New Haven who says he has left Prof. Marsh's employ," he wrote to Henry Fairfield Osborn, and told him of the allegations about Marsh having plagiarized the Dinocerata monograph that Williston would publish in Oscar Harger's obituary.

Edward the sorcerer had found a good familiar in Osborn, who resembled Marsh in financial and political acumen and in his preference for scientific administration over field and lab work. With a wealth of connections among his father's railroad tycoon associates, Osborn was able to play a role in the later bone war similar to Marsh's in the Powell-Hayden struggle, by marshaling anti-Othniel forces behind the scenes and writing damaging letters to legislators. Still a cloistered academic in 1885, he was less aggressive than Cope would have liked, but Marsh would find him a deadly opponent as his power grew during the next decade. Indeed, he would prove better at the game than Marsh, because he had a cooler head and never risked his money or reputation.

Cope, who didn't share his young ally's prudence, was ready to rush in for the kill. A letter to Annie during his September Yellowstone trip showed him indulging in daydreams. "As to the Washington position," he wrote, "I have little expectation of its being offered to me, but if it were I would accept it in self-defense. I would reorganize the survey as to men, take the position of Vert. Paleontologist, and turn over the chief position to someone else, on whom I could rely." Perhaps he imagined Marsh and Powell boiling in the geyser basins, about as likely as a Cope-led U.S.G.S. takeover.

The daydreams may have been idle, but he came up with a more realistic strategy in October. Cope would use the testimony of disgruntled survey personnel to discredit it, and he got his mining engineer, Frederic Endlich, to canvass Powell's men for damaging items. "I presume you are aware of the fact that the Powell Survey is going to the wall," an Endlich letter read. "I want to know all about the deadheads on the survey, favoritism, misapplication of funds, waste of money, etc. If you are in a position to give me the information, I shall be very much obliged and will remember it in the sweet bye and bye." Even though wise subordinates simply turned the letters over to Powell, some of the mud shook loose. Representative Herbert began asking unexpected questions at hearings, mostly about things the survey had done under Clarence King's lax supervision. Herbert had found at least one crack in the establishment. Having made a private fortune in mining, Cope's old enemy Alexander Agassiz saw no need for government science except where it benefited his specialty, marine biology. Partial to the Coast Survey, he obligingly called Powell's survey wasteful and extravagant when Herbert called him to testify.

The seething Peabody lab offered an even richer source of mud. Cope wrote to Osborn in October that Williston and other assistants who'd left Marsh's lab wanted to "place themselves and him right before the Scientific public. They have found M to be more of a pretender than I had supposed him to be . . . It is now clear to me that Marsh is simply a scientifico-political adventurer who has succeeded in ways other than those proceeding from scientific merit, in placing himself in the leading scientific position in the country . . . This winter there is to be a thorough ventilation of the Geological Survey at Washington. Marsh will probably try to get the National Academy to boost the organization. There is a good deal of Marshiana in other departments than the Vert. Pal. as I am told. In any case, such an opportunity of placing Marsh in his true position will not occur again."

Cope wanted the defectors to publish an anti-Marsh manifesto, but for some reason this wasn't possible without George Baur's participation. Edward had been playing Cassius to Baur's Brutus—he named a dinosaur after the young German and offered other blandishments—but Baur remained employed by, and indebted to, Marsh. Cope asked Osborn to lend Baur the money to pay Marsh off, but Osborn had more sense, and the manifesto languished, despite Cope's enthusiasm. "It should, if possible, be published in *Science;* but, failing that, separately," he wrote to Osborn on November 28. "I have told the boys that I should have no connection with it in any way . . . A newspaper man of New York tells me that he will see that it is well circulated in the Press! It is a business I do not like, but it is absolutely necessary, or our Government aid to Science will be forever ruined and the reputation of the country compromised."

Cope's excitement was premature. The most he accomplished during the 1884–1886 investigation was to circulate in Congress a 23,000-word anti-U.S.G.S. report that he, Endlich, and Persifor Frazer had compiled. This helped Representative Herbert come up with his unexpected questions, but Powell was much too good an administrator to succumb to confidential slurs and hearing-room surprises. Cope's allegations didn't make the highbrow papers, much less the sensationalist ones, which were preoccupied with recent allegations of cannibalism during a polar expedition. "It was somewhat unfortunate that Powell was allied with Marsh, for he was sooner or later to have his flank exposed by Marsh's intemperate feuds," Stegner wrote. "But for the time being, at least, and thanks mainly to Powell's testimony, government science and especially the geological survey came out of the commission hearing in 1886 much strengthened . . . The charges circulated by Cope and Endlich were not read into the record of the testimony, and the spies and whisperers slipped back behind the arras to await another chance."

Cope kept busy playing fox in the Marsh coop. "In a few days I will leave here," he wrote to Annie in July 1886, from his endless wait in Interior Secretary Lamar's anteroom. "I want to stop over a day in New Haven and see Dr. Baur." In August, he told Osborn that he had spent a day at New Haven with Baur and Williston. Schuchert recalled that he had stayed with Baur in New Haven during an 1888 National Academy of Sciences meeting. He even may have haunted Yale in the guise of a football fan.

Yet Cope's intrigues accomplished so little that Marsh's outrage seems overwrought, and another attempt to topple the ruling geological junta in 1887 did even less. Again, Cope was enthusiastic in a letter to Osborn. "It is now I think a favorable time to make an effective move toward terminating the 'iniquity,' as you have called it," he wrote on November 21, "and I think it will require only one more witness besides Williston, Meyer, and myself, one who is neither an employee nor a rival for position, to make the whole thing conclusive . . . I will be only too glad of anything you may choose to write in confirmation of what I have said to Mr. Page, U.S. Treasurer. I told him . . . of the orders to break up fossils, and of the employment of guides to obstruct your work. If I can get confirmation from you, I think it will be the beginning of the end of O.C.M. in Washington—or at least the beginning of my ability to get my work finished and published."

Again, however, Osborn declined to become as deeply involved in the feud as his impoverished friend wanted, and Edward sounded discouraged in a November 29 letter to his daughter. "I have been at work here for nearly a week and can't say I have made very satisfactory progress," he wrote from Washington. "Marsh appears to control the National Museum as he does the Geological Survey and it is one of the most disgraceful facts of the present time." A diminution of zeal among Cope allies as showdowns loomed was one of the feud's constants. They had their own problems, of course, and, unlike Osborn, weren't independently wealthy. Edward had just one ally whose eagerness for battle matched his own—the "newspaper man of New York" he'd mentioned in his November 1885 letter to Osborn.

William Hosea Ballou said he'd become "a close and fascinated reader" of *The American Naturalist* while he was in college, and, though fourteen years younger than Cope, had been the naturalist's "intimate friend" since their first meeting after he graduated. They were, he said, "so intimate that we even indulged once in the most violent quarrel and the next day forgot it." In a popular magazine, *The Chautauquan,* he claimed that he'd been "the first, last, and almost the only journalist" through whom Cope's "discoveries were given out to the public press." Ballou had some writing ability. Alfred S. Romer recalled that "some of his work, at least in the early period, was respectable popular science." In the 1870s and 1880s, he published competent articles in *The American Naturalist,* including a "Proposal for an Adirondack National Park," marred only by a claim that the Hudson was "perhaps the oldest river in existence."

Ballou's *Chautauquan* article suggested he understood little about post-Cuvierian paleontology, however. "'Ancestor,' by the way, means only the highest type of animal in the preceding geological age," he wrote. "'Evolution' amounts to that and little else.

Man could not have existed in any of the geological ages of his 'ancestors' as there were not conditions of soil and atmosphere to support him. He will have no successor, because divine authority assures him he was created in the divine image." Romer knew him as a writer of "pseudo-science for the Sunday supplements, with articles like 'What was the animal whose eyes the tropical explorer saw gleaming in the depths of the cave in the jungle? Was it the gigantic *Brontosaurus* or the ferocious *Tyrannosaurus*?' "

Ballou's sensationalism evidently touched a chord in Cope's sometimes childlike sense of wonder. In one letter to his daughter, Edward described a wild goose chase after a "sea serpent" mentioned in a newspaper. Ballou had "said the sea serpent was at a taxidermist's nearby; so I went with him. We visited two of these gentlemen and only one of them had seen it in a barrel, and it was sent away, he did not know where!" Edward probably also enjoyed the younger man's hero worship. "Can we pass him by as merely the greatest naturalist of his time," Ballou queried, "when such men as Darwin, Agassiz, Linnaeus, Humboldt etc., stand more or less discredited in the light of larger collections and more exact data? I am of the opinion, to say the least, that in the dim future . . . the name of Cope will be the illustrious one and that he will form most of the base of the American scientific pyramid."

Yet he was not an ally to inspire much confidence, even in someone as vain and self-absorbed as Cope. His self-aggrandizing *Chautauquan* article belied a certain disdain on Edward's part. "I rushed over to see him every time I recognized news value in his work," Ballou wrote. "In working hours he would stop to talk and explain at intervals, and then order me to 'shut up' and 'not stir,' while he wrote out the descriptions running in his head of the new-found bones. When he got his view of a bone on paper, he would turn around and say, 'Well, where are we at?' Then for from a few minutes up to two hours he would pitch into the news story of his particular discovery I was endeavoring to draw out of him. 'I wouldn't waste my time on you,' he would reiterate, 'if your queries didn't help me round out my own view of things and offer suggestions for further examination.' Then again: 'Newspaper men have some value after all. They keep a man from getting as rusty as a bone he is working on.'"

If Cope looked down on Ballou's scientific ignorance, Ballou considered Cope a journalistic innocent. "He never had the news sense developed to the slightest extent," he wrote. "Cope was bitterly assailed by a coterie of government scientists . . . He could not get justice in Washington, nor his works printed, nor his pay from the government. I undertook to straighten things out for him, at which he was greatly amused. 'Why, little boy,' he said, 'you don't want to get in-between. They will make dust of you so fine they can't even see it on their feet.' 'I'll show you,' said I, 'that the press rules in this country.'"

Despite Ballou's pipsqueak boasts, 1889 must have seemed like the past years to Cope. He was still sliding into penury and flailing impotently at Marsh. "I don't know where my next month's board is coming from," he told W. B. Scott. Not only did he lack ready cash to pay his taxes and mortgage; he was still in debt for *The American Naturalist.* "I have endless trouble with these publishers and have secured now a new firm which will take it up at once," he wrote to Osborn. "But I owe for editorial work for 1888–9 $455.00 which I cannot pay, having raised only $45 out of the $500 for that purpose. Can you give me a lift with the difficulty? Sorry to have to ask."

In April, there seemed to be some hope of deposing Marsh from his National Academy of Sciences presidency. Cope and his partisans tried to persuade others to join them on the grounds that Marsh had allied himself with physicists and chemists to rule at the expense of biologists and geologists. According to Robert Plate, "Marsh had been in so long that the Academy was ready to elect someone else, but Cope's strenuous attacks backfired. His onslaught was so extreme that the members felt obliged to defend Marsh, and themselves as well, for having elected him." Whatever the reason, Marsh won easily. "I see the National Academy has disgraced itself again," Cope told Osborn. "The vote was 22 to 13. I know of 5 or 6 men who remained away, who wd. have voted against him. Such is the cowardice of a certain specimen of alleged man."

He still didn't give up. "I started the recommendations for yourself and Scott for membership for election next April," his letter continued. "The list of members is now full (100) but there will probably be two vacancies by April." At the rate of two partisans a year, Cope could hope to pack the academy in his favor by around 1900, but he was ready to try. "I hear that Marsh is to resign soon from Presidency of the National Academy, and that he sought a re-election as a 'vindication,'" he wrote to Julia on April 29. "Why he thinks he needs a 'vindication' is what I do not understand, and why the sage physicists etc. should want him to have it is stranger still."

Evidently 1889 was a year for strange things. Just when Cope seemed on the brink of plunging forever below any hope of a professional position, he was lifted mysteriously back into the light. Although no documented source offers an explanation, the University of Pennsylvania suddenly appointed the contentious virtuoso to a professorship for that fall. Someone apparently had been working to end the unseemly spectacle of a major American naturalist hung out to dry for a decade, but it's unclear who. It may have been Cope's young academic friends, though Osborn, usually candid about his generosities to Edward, took no credit. Perhaps Joseph Leidy and other Pennsylvania faculty were the prime movers, though Cope thought they were against him, and W. B. Scott wrote that "he was extremely unpopular in Philadelphia, and there was never a better illustration of a prophet without honor in his own country."

The mysterious professorship was not very lucrative, but it did put Cope back on his feet, providing his first steady income in years. It was as though the sensationalist fate that had been bouncing him and Marsh against each other since 1863 was at it again, reviving the loser when he finally seemed down for the count. If so, it had what Gore Vidal calls "a sense of fun." Cope's new professorship was not in paleontology or zoology, but in geology and mineralogy, O. C. Marsh's sadly relinquished boyhood passions. Fun-loving fate was perceptive, too. A man other than Cope might have been subdued by all those years of knockdowns and might have tried to avoid further trouble after regaining his feet. A man other than Marsh might have felt magnanimous after decking a rival so consistently for ten years that a minor academic post came as a lifesaver. But sportive fate knew its men.

14

Cope Strikes

One day in late 1889, Professor Cope got a letter from Secretary of the Interior John W. Noble, only recently appointed. Dated December 16, it ordered him to place his fossil collections in the U.S. National Museum on the grounds that they were government property, Cope having acquired them while working with the Hayden survey. The "lively scene" that ensued on Pine Street can be imagined. Cope had spent at least $75,000 of his own money on the fossils, and they were all that remained of his capital. Anxiety about their ownership had been tormenting him at least since 1878, when he'd written to Ferdinand Hayden that he'd returned all the specimens he should have, and that "O. C. Marsh has had more assistance from the Govt. than I have had—and his collections go to Yale College!" Noble's letter assumed, moreover, that Cope's surrender of his fossils to the government was a preliminary for publishing the rest of his Tertiary mammals "Bible," something Powell was unlikely to do, as Edward knew from bitter experience.

"Secretary Noble had ignored the facts on the recommendation of Director Powell," wrote Osborn drily. In his turn, Powell probably had "ignored the facts" on the recommendation of Professor Marsh. What the two manipulators hoped to accomplish by this breathtakingly unjust ploy is not clear. Powell may have wished to throttle Cope's tiresome attempts to get his "New Testament" funded, and Marsh may have wanted revenge for Cope's latest campaign to depose him from the academy presidency. According to Url Lanham, "the move was pure malice, perhaps made with the knowledge that Cope would be able to defend himself against the manifestly unfair claim, but only after producing a good deal of fun for his tormentors." It seems doubtful that Marsh and Powell enjoyed what followed, however.

Secretary Noble might as well have flung a torch into the Washington arsenal. The stored ammunition went off promptly, although the explosion's exact origin is unclear. Osborn thought William Hosea Ballou "persuaded Cope that now was the time to strike," but it is unlikely that Ballou had that much influence. Cope had been dying to publish his Marshiana for years, and only his friends' lack of cooperation had restrained him. Now he had a pretext. "I don't think writing private letters of a critical kind to such hardened sinners as Marsh and Powell does the least good," he would write to Osborn after the fireworks began, "especially in the case of Marsh. I tried charity and forebearance with him for years but he is impervious to such agency. He is thoroughly case-hardened. But when a wrong is to be righted, the press is the best and most Christian medium of doing it. It replaces the old-time shot gun and bludgeon and is a great improvement. I do not share in the objection to newspaper exposures in such a matter."

Getting the "shot gun and bludgeon" replacement into the papers was not as automatic as Ballou made it seem when he later bragged of his "fifty-two-column exposé running eight days in the *New York Herald.*" Probably influenced by such boasts, Osborn called him "a young newspaper man who worked for the *New York Herald,*" while Alfred S. Romer made him "a journalist on the staff with some acquaintance with science," and Wallace Stegner promoted him to a full-fledged "*Herald* reporter." Osborn even thought he wrote headlines, though it would be a rare major daily that let a reporter do so. The historian Elizabeth Noble Shor showed Ballou to be an unreliable source, however. When she carried out research

William Hosea Ballou in 1910

on his 1938 *Who's Who* entry, she found most of the entries to be unverifiable or false.

Ballou may have been descended from an illustrious family. Others with his name were the first American Universalist clergyman, the first president of Tufts College, and the founder of the *Boston Globe.* If so, he was a black sheep. He said he had attended or been granted honorary degrees from five colleges and universities and had served on four government surveys and expeditions, but Shor could find no convincing documentation for these claims. She determined that he had published some fiction along with his journalism, and had contributed specimens to a Smithsonian fungus collection, but decided that his *Who's Who* fabrications and some strange ideas (he thought a fish fungus caused human cancer) put him on the lunatic fringe. "After removing the unconfirmed and unconfirmable items in the above account," she concluded: "Who *was* this . . . ? Chiefly he appears to have been a novelist, a free-lance writer and popularizer of science, and an amateur mushroom collector, with delusions of grandeur." Indeed, a photo of Ballou, showing a sloping forehead, receding chin, shifty eyes, and strangely convoluted ears, might have come from the period's abnormal psychology textbooks.

One claim Ballou did *not* make in the 1938 *Who's Who* was that he ever had been on the staff of James Gordon Bennett, Jr., and Shor concluded that he "was neither a reporter nor a correspondent for the *New York Herald.*" Other evidence supports this. Even Ballou himself once let slip his low journalistic status in a letter to C. D. Walcott, a paleontologist. "I shall not be held responsible for sensational headlines, captions on pictures, and introductions by editors," he complained. "I can't help myself, can I, if I deal with layman editors?"

W. B. Scott wrote that a few weeks before the affair got into the *Herald,* Ballou had called on him with an introduction from Cope. "I had heard of him before and knew that he was a friend of Cope's, who had named a fossil after him," Scott recalled. "When, therefore, he told me that he was investigating the U.S. Geological Survey, in general, and the scientific iniquities of Powell and Marsh, in particular, I had complete confidence in him and told him what I had heard . . . Nothing was said about newspaper publication and I had no suspicion that the caller was a newspaperman and meant to print my remarks." That wasn't surprising, since Ballou wasn't a newspaperman and hadn't found a publisher for Cope's Marshiana.

Finding one wasn't easy. A *New York Times* article revealed that Ballou had peddled the Cope-Marsh material around to various newspaper offices during Christmas week of 1889—hardly a staff reporter's procedure—and that the *Herald* was the only paper that would buy it. His position in *that* deal would have been a particularly abject one. The *Herald* was not just any major daily, of course. It was James Gordon Bennett Jr.'s private totalitarian dictatorship.

Bennett's despotism over the *Herald* perhaps was not quite as absolute in 1890 as it had been earlier, because he hadn't lived in New York for twelve years. Still, his zombies enjoyed little freedom. The Commodore had helped to create the transatlantic cable, and he used it on his paper like a bullwhip. Often working from dawn to dark, he held sway via what Richard O'Connor described as an endless "two-way stream of memos and cables." An executive was stationed at the New York end of the cable at all times, and Bennett's empty office was maintained with sharpened pencils and a fire in the grate. Bennett also maintained a network of spies, called "white mice," who informed on the staff's personal as well as professional lives. Periodically, he would cross the ocean and descend with "terrifying suddenness" on the paper, firing anyone who vexed him.

It seems unlikely that a story meriting as much space as the bone war could have escaped Bennett's notice, and even more unlikely that a free-lance hack like Ballou would have been given control over it. Once the *Herald* decided to use the feud, the zombies probably paid off Ballou and brusquely showed him the door. He got a byline because he was *not* on staff. The only "staff" name the paper printed in 1890 was "James Gordon Bennett, Proprietor." Editors even had to sign the Commodore's name to their correspondence.

Bennett's reason for devoting so much space to a fossil feud remains as unfathomable as the rest of his behavior, but one circumstance conspicuously worked in its favor. In November 1889, Nellie Bly, the *New York World*'s wildly popular stunt reporter, had set out to circumnavigate the globe in less than the eighty days it had taken Jules Verne's fictional Phileas Fogg. It was the newspaper sensation of the year, and Bly would return to a hero's welcome in mid-January, having accomplished the journey in seventy-two days. The *World*'s appropriation of the sensationalist market had been a thorn in Bennett's side ever since the paper had opened, in 1883. He'd tried to crush it with a price war, but that had backfired, and in 1890 its circulation surpassed that of the *Herald* for the first time. The Commodore would be damned in hell before he'd cheer Joseph Pulitzer's stunts from the sidelines, so the *Herald* conspicuously ignored "globetrotting Nellie Bly." It needed something to screen the eloquent jealousy of its silence, however, and the bone war was handy.

Indeed, Cope's attack on the geological survey must have seemed heaven-sent, judging from the enthusiasm with which Bennett's paper initially embraced it. The *Herald* went after Powell and Marsh the way a naval destroyer attacks submarines, a strategy suited, albeit anachronistically, to the Commodore's nautical bent. It dropped a blunt salvo of depth charges—Cope's and Ballou's Marshiana—and watched to see if any wreckage rose to the surface. Wreckage quickly did, even before publication.

"Some weeks later," W. B. Scott recalled, "I was greatly surprised to receive a very curt and peremptory letter from Marsh, saying that the *New York Herald* was threatening to print an attack upon him by Cope and others, that he had seen the article, which contained certain derogatory statements by me, and demanding to know whether I had authorized the publication of these statements. I immediately replied that I had authorized no publication and did not even know what I was alleged to have said; I also wrote the editor of the *Herald* asking him to suppress anything that had come from me. Though I thoroughly disapproved of Marsh and would have rejoiced to see him removed from any connection with the U.S. Survey, I did not like at all this sensational method of newspaper attack. The editor of the *Herald* declared that my letter had not been received, a perfectly incredible statement . . . Had I been better versed in slippery ways, I should have registered the letter and got a receipt for it."

Scott also described another bit of prepublication wreckage: "Marsh showed great ability in attempting to ward off the threatened attack and was not very scrupulous in his methods of attack." Learning that Provost Dr. William Pepper of the University of Pennsylvania was "involved in a blackmailing case, the rights and wrongs of which I never knew," Marsh wrote to the unfortunate man, "demanding that he silence Cope on pain of having his own scandals aired." Pepper took the demand to heart. "Marsh has made a dead set at Pepper . . . so as to secure my resignation or expulsion," Cope wrote to Osborn a week before the first article appeared. "Pepper is terribly frightened and yields everything to him . . . It will require considerable effort to prevent my being 'retired' from the University. Now it is assumed by Pepper that my statements as to Marsh are false, the result of personal spite . . . What I need is some credible person to inform Pepper as to the facts." What Cope really needed, in fact, was someone on his side who would frighten Pepper even more than Marsh had, and there was someone who did. "The *Herald*, getting wind of this," concluded Scott, "let the Provost know that he would have cause to regret any attempt to interfere with Cope's freedom of speech."

The bone-war articles included a further sign of journalistic enthusiasm when they started coming out on Sunday, January 12. Lengthy editorials ran with them. Osborn and later writers largely ignored these, but they were among the most striking aspects of the *Herald*'s coverage, and W. H. Ballou was even less likely to have written them than headlines. Although the *Herald* was not famous for it, there was always a daily editorial page. According to Bennett's biographers, editorials ran with his knowledge and often at his insistence, as when he cabled from Ceylon once to order a piece on a New York matter. Their labored style showed his influence, and he sometimes wrote or edited pieces himself.

The editorial that ran with the first feud feature labored hard to give the affair the dizziest possible spin. Headed A GREAT SCANDAL AMONG SCIENTIFIC MEN, it began, "Very serious trouble in connection with the Department of Geological Survey is brewing. Rumors of War have been heard for several years, but the first Krupp gun is fired in the *Herald* this morning." Readers were to understand straightaway that Bennett's paper was giving the run of its columns not to a mere scientific squabble but to the prospect of government upheaval on the order of the Red Cloud affair.

"We have a country nearly as large as the whole of Europe," the editorial proclaimed. "Its mining resources have been developed to such an extent that we take out of the earth annually something like six hundred million dollars. But the future bids fair to outdo the past. Untold wealth is waiting for the enterprise of pioneers; millions are slumbering undiscovered. The business of the Geological Survey is, in part, to search for the hiding place of precious minerals, to indicate the area in states and territories that might profitably be worked. For this and other allied purposes Congress appropriates an immense sum of money every year, and at the head of the Survey is a gentleman who is supposed to be impervious to corrupt influences, who will not favor rich mining companies, who is in no degree liable to monetary temptation."

This was an imposing fanfare, although it was mostly bluster and displayed some ignorance of the main issues pertinent to the U.S.G.S. Precious-metal mining was not one, since its congressional enemies had restricted the survey to the public domain, from which mining interests had hastened to remove lucrative deposits. The editorial erected a straw man by alleging that John Wesley Powell, a reformer, might be tempted by "rich mining companies." Not even Powell's worst enemies believed that. Accuracy, though, was not the point. The public might believe such insinuations, knowing even less about the U.S. Geological Survey and its chief than did the editorial writer, and it was the public that the *Herald* was after. Bennett "liked the novel and bizarre and he did not mind if people ridiculed him and the *Herald*," wrote Oswald Garrison Villard, editor of *The Nation*. "What he dreaded was them not talking about his papers."

"We give a good deal of space to the quarrel for two reasons," the editorial continued. "First—it is a matter of the utmost importance that the Geological Survey should be manned by scholars of undoubted competency and unimpeachable honesty. Second—it is equally important that when acknowledged specialists like Professor Cope openly make grave charges against such gentlemen as Major Powell, Director of the Survey, and Professor Marsh of Yale College, his chief assistant, to give testimony on both sides in full." It then shifted from solemn bluster to florid martial rhetoric: "Professor Cope, of the University of Pennsylvania, uses very strong English. He has instituted war to the knife and proposes to bury the blade to the hilt in the heart of his adversary. He is defiant, disdains innuendo, and deals in plain assertions. His assault is bold to the verge of recklessness and vigorous to the edge of audacity. He is either everlastingly right or infamously wrong. No compromise is possible. Some one ought to lose his reputation as a result of this fight."

Then, reverting to bluster, the editorial concluded, "It is a serious and grave affair in which a number of valuable reputations are involved. Professor Marsh is quite able, no doubt, to defend his scholarship. That, so far as the public is concerned, is a matter of great interest. Major Powell may be able to prove that he has in no wise been false to his great trust. That is a matter of serious importance to the people of this country. Professor Cope nows what he is about when he makes these charges, and should be well fortified with facts. Gentlemen, please to remember that criminations and recriminations are not argument. The 'you are another' policy is barred. The main question is this: Is the United States Geological Survey conducted by the best man in the best manner? If it is, Professor Cope will retire to the background. If it is not, Major Powell and Professor Marsh must answer to the people."

Words calculated to stir the reader, and a whiskered gentleman reading them in some ornate Manhattan breakfast room might have turned to the article with anticipation. Those scientists and government men, gad! As he found the page, his eyes would have been blandished by Jovian line drawings of the whiskered antagonists. The 1890 Sunday *Herald* was not as lavishly designed and illustrated as Sunday papers soon would be, but it was headed in that crowd-pleasing direction. The headlines would not have disappointed him. They were piled in multiple decks down the first column, a standard attention-grabber before Pulitzer and Hearst popularized the banner:

SCIENTISTS WAGE BITTER WARFARE
Prof. Cope of the University of Pennsylvania brings Serious Charges
Against Director Powell and
Prof. Marsh of the Geological Survey

CORROBORATION IN PLENTY

Learned Men Come to the Pennsylvanian's Support with Allegations of Ignorance Plagiarism and Incompetence Against The Accused Officials

IMPORTANT COLLATERAL ISSUE

The National Academy of Sciences, of Which Professor Marsh is President, Is Charged With Being Packed in the Interests of the Survey

RED HOT DENIALS PUT FORTH

Heavy Blows Dealt in Attack and Defense and Lots of Hard Nuts Provided for Scientific Digestion

WILL CONGRESS INVESTIGATE?

The opening paragraphs also would have been promising, even though repetitious. "For some time past a volcano has been slumbering under the Geological Survey, and of late there have been indications that the time for eruption is not far distant," the *Herald* declared, attributing the "commotion" to "the dissatisfaction with which Professor Cope and some of his professional associates in the field of vertebrate paleontological research have long regarded the conduct of the survey under Director Powell and Professor Marsh." The whiskered gentleman's eyes would have glazed at "vertebrate paleontological research," but would have brightened at the next heading.

A BIG POLITICAL MACHINE

"The charges which are now brought forward by Cope and his friends tell the whole story," the feature continued. "They cover a great deal of ground and touch on every branch of the subject."

The *Herald* promised to reveal how, according to Cope, Powell and Marsh had conspired to pervert the Geological Survey into a "gigantic politico-scientific monopoly run on machine political methods," grab control of the National Academy of Sciences, and commit other crimes, including plagiarism of scientific works, sequestration of government collections, falsification of survey reports, and bribery of politicians, scientists, and journalists "in order to disarm them of any feelings of hostility they might entertain toward Director Powell and the institution under his charge."

It wasn't Stanley in Africa, the whiskered gentleman would have reflected, but it might be worth looking into. As he read on into the section headed PROFESSOR COPE'S CHARGES, however, his eyes would have glazed again. It began with a long list of Powell's and Marsh's scientific enemies, including their degrees and academic positions. That might be "corroboration in plenty," but what about the "hard blows"? Well, skip that paragraph. Next came descriptions of the Geological Survey and National Academy of Sciences that echoed the introductory section at length. Skip that too. Next came repetition and elaboration, in technical detail, of the charges already made against Powell and Marsh. Ho hum.

The whiskered gentleman's eyelids would have lifted once more at the next heading: PROFESSOR COPE WOULD NOT BE SUPPRESSED. The editorial had called Cope "bold to the point of recklessness." Let's see what the fellow had to say. "The scandals within the Geological Survey might have been kept out of print for some time longer perhaps if Director Powell had not attempted to suppress Professor E. D. Cope. When Professor Cope felt himself the victim of an outrage, he carried his case before the Secretary of the Interior." What outrage? What case? Lured by another come-hither head, PROFESSOR COPE SPEAKS HIS MIND, the whiskered gentleman would have pushed through yet another dense paragraph of scientific credentials and then plunged into the relative clarity of an "interview."

Now would the recklessness start? The interview began briskly enough with Cope "engaged in naming new species of fossil animals" at his Philadelphia home. If the whiskered gentleman had been a naturalist, he would have read on with interest, since Cope used the interview to savage Marsh's and Powell's scientific reputations. He said that Marsh had plagiarized his evolutionary work on horses, toothed birds, *Dinocerata,* and horned dinosaurs, and that he had mistaken a horned dinosaur for a bison and a dinosaur for a toothed bird. He said that Powell was "ignorant of the best-known facts of geology" and had committed all kinds of scientific blunders in running the survey. The whiskered gentleman's outdoor interests probably stopped at a bit of rough shooting, however, and he might have given the paper a shake. This was "vigorous to the point of audacity"? These were "stunning revelations of incompetence and malfeasance"?

Accusations of malfeasance were, in fact, scattered among Cope's replies to formulaic questions, but not the kind that would outrage whiskered gentlemen. Marsh signed his name to his assistants' work. My, my. Powell gave government jobs to scientists and journalists he liked and not to others. What else was new? "Scientific men usually are very simple of thought," Cope said, "unsuspecting of the machinations of others, and thus have been the victims of a pair of political scientists, more political than scientific." Simple indeed, the whiskered gentleman might have muttered.

Cope's recklessness then stopped as abruptly as it had begun. "As this interview seems to be at an end or at least at a point where confirmation will be called for," he said, "I will place in your hands all the correspondence in my possession. From it you can extract a sufficient justification for all that I have said." The whiskered gentleman might have felt he'd been led pretty far down the garden path. Bother paleontologists, time for a trip to the lavatory, perhaps a stroll on Fifth Avenue. On the other hand, he'd already read half the article. There had to be *something* in it. The next item, a letter from a Professor Williston at Yale, might have brightened his eye again. Williston got down to cases about Professor Marsh.

"I never knew him to do two consecutive honest days' work," he said, "nor am I exaggerating when I say that he has not averaged more than one hour's work per day . . . Those who know him best say—and I concur in that opinion—that he has never been known to tell the truth when a falsehood would serve the purpose as well."

Yet Williston's fulminations ended soon enough, and if the whiskered gentleman finally dropped the paper and took his stroll, he would have had just cause. A letter from the former Peabody Museum assistant Otto Meyer emphasized Marsh's laziness and dishonesty; three long letters between Cope and Interior Secretary Noble rehashed the subject of Powell's fossil-grabbing; and so on. "Sufficient justification" reached a crescendo of dullness as Persifor Frazer attacked Powell's "efforts to destroy the American Committee of the International Geological Congress, because he was not permitted to control it," a matter as immaterial to Cope's accusations as it was to *Herald* readers. The article then dribbled away with non sequiturs by Eugene Smith, an Alabama state geologist, who had "nothing to complain of."

William Hosea Ballou's byline followed, but most *Herald* readers would have joined the whiskered gentleman by then, and no wonder. Ballou's piece was among the more feeble bits of journalism published by a major daily, an awkward, boring patchwork. It wasn't even authentic, since Ballou had faked his interview with Cope and misquoted other people he claimed to have interviewed. SUPPRESS IT? NOT MUCH! read a subhead under the byline. The item that followed concerned Marsh's attempt to muzzle his detractors, with Ballou defending his unauthorized use of Scott's and Williston's comments. But that subhead had an ironic ring. If anything could suppress Cope's charges against Marsh and Powell, it was Ballou's dullness. "Suppress it? Please!" would have been the whiskered gentleman's response.

It was not much of a start for a sensational scandal, and the editorial cheerleading that accompanied Ballou's article showed that the *Herald* knew it. To be sure, there had been the gratifying wreckage of Marsh's attempt to suborn Dr. Pepper, but dropping depth charges has its dangers. Sometimes it's torpedoes that emerge from the depths, not wreckage, and the charges dropped on John Wesley Powell had elicited a swift and formidable response, which the *Herald* could not in good conscience have failed to print.

Delivered with what Wallace Stegner aptly called "restraint and decorum," Powell's counterattack was as powerful a piece of journalism as Ballou's was feeble. After graceful bows to the President, the secretary of the interior, Congress, fellow scientists, the American people, and "newspapers throughout the country . . . the guardians of official integrity," Powell wasted no words in carrying the fight into his enemies' camp. "What is the origin of this war against the Geological Survey?" he asked. "The question has a simple and perfect answer." Cope and his allies wanted to appropriate it, and had been trying to do so for years. "During all this time, these men have not failed to secretly stab at me and my assistants whenever opportunity could be found," Powell claimed, and briskly marshaled the evidence: including Cope's 1884 "scotch Powell" memo, Endlich's 1885 mud-searching letters, and the 23,000-word attack circulated to Congress during the 1885 investigation. Because the attack hadn't been read into the congressional record, he concluded, Congress had found it "baseless," and that itself amounted to "a complete refutation of all the charges made against the Geological Survey and its administration."

"Having put Cope where he belonged, in an envious minority," as Stegner put it, Powell proceeded to dodge the charges of political manipulation with some convincing, if not entirely candid, rhetoric. He praised Marsh and denied that the Professor had mistreated his assistants—at least not his survey ones—or sequestered specimens, but he also distanced himself subtly from his quarrelsome ally. He'd had, he wrote, "only a brief acquaintance" with Marsh at the time the Academy of Sciences was making its 1879 recommendations, and he had later chosen him as survey paleontologist because Leidy didn't want the job. Powell made it sound, indeed, as if the survey directorship had fallen into his lap by accident. "I did not expect to become connected with it when it was organized," he wrote. "In 1881, Clarence King resigned in order to engage in great mining enterprises, and my nomination as director of the Survey was therefore sent to the Senate without any solicitation and without my knowledge."

After delicately touching up the survey's past, Powell painted a glowing picture of its present. "It has been charged against me, as if it were a crime," he proclaimed, "that the members of the Academy countenance and support the Survey under my charge; that the professors of the colleges and universities of the land are my friends and sanction my work; that the great body of geologists of America sympathize with my purposes, and that even the Senators and members of Congress of the United States give me their support; that all of these are in league with me in some scandalous affair. And what is it, forsooth, that all these men have combined to accomplish which is so wicked? Nothing less than to put under a bushel the light of the glory of the genius of Professor Cope."

Furthermore, he hadn't tried to put Cope under a bushel. Cope himself had sabotaged the publication of his second volume on Tertiary vertebrates by failing to submit the whole manuscript and refusing to turn over his fossil collection to the National Museum, as Secretary Noble had ordered. Cope was "more interested in the description of species and the publication of his name as a label thereto than in the profounder problems of the science of paleontology . . . In his determination of geological horizons or the age and succession of rock formations, which is so important to a geological survey, he is careless and unsound." This was getting a little technical and wide of the mark, but Powell adjusted his aim and concluded with one of the more masterly *ad hominem* indictments in American newspaper history. Its subtle shadings of blame and praise put Cope on trial, weighing the paleontologist in a balance between professional perdition and redemption.

"For all these reasons, I believe that it would be incompatible with the trust imposed upon me to employ Professor Cope on the survey," Powell declared, "though at one time, until the facts came fully to my knowledge, I was intending to employ his services. Moreover, the paper to which I am now replying is sufficient evidence to any fair-minded man that Professor Cope's mental and moral characteristics unfit him for any position of trust and responsibility. In addition to his great vanity, which leads him into vicious species work, he is inordinately jealous and suspicious of every other worker, and these two traits combined give him that hysterical temper and gift of voluble denunciation rarely found in persons of his sex. In fact, his general ravings about scientific men, members of the National Academy, professors in college, and geologists in general, whom he believes are all in league against himself, make it impossible for him to associate on terms of cooperation with other men engaged in kindred work."

The *Herald* rudely interposed the subhead NO INJUSTICE TO THE PROFESSOR INTENDED at this point, but Powell summed up judiciously: "I am not willing to be betrayed into any statement which will do injustice to Professor Cope. He is the only one of the coterie who has scientific standing. The others are simply his tools and act on his inspiration. The Professor himself has done much valuable work for science. He has made great collections in the field and has described these collections with skill. Altogether he is a fair systematist. If his infirmities of character could be corrected by advancing age, if he could be made to realize that the enemy which he sees forever haunting him as a ghost is himself, and if he could be made to see that it is of importance that he should promptly fulfill his engagements with other men, he could still do great work for science."

Powell's torpedo might have blown the affair out of the water if it had caught the public's attention. But he should have known better than to send it to the *Herald*. The paper dodged it by burying it after Ballou's interminable screed, surely aware that few readers would get to it. *Herald* readers often didn't finish even the most sensational articles, as the 1874 Central Park Zoo hoax had demonstrated. The last paragraph of that account had admitted that the wild animal stampede was "simply a fancy picture which crowded upon the mind of the writer a few days ago when he was gazing through the iron bars of the cages," but many readers absorbed just enough about "carnivorous beasts, [and] their lurking places not known for a certainty," to panic. Some had rushed to the park with guns.

Steering clear of Powell's torpedo, the January 12 *Herald* concluded by displaying some of its wreckage. A final head read: THEY ALL DENY: DENIALS WHICH SOUND FUNNY IN THE LIGHT OF PREVIOUS EVENTS. The disclaimers Marsh had extorted from Osborn, Williston, Baur, Scott, and John Bell Hatcher followed verbatim, and they did sound funny. "Just returned from the South. Have not seen or authorized any article whatsoever," wrote Osborn. "I hereby voluntarily state," wrote Baur, "that I have never in any way authorized the use of my name in any attack on your work." The vituperative Williston made two disavowals: "The whole subject no longer concerns me, and is distasteful"; then, two days later, he added that he hadn't authorized Cope or any other person to attack the Yale Professor, and that Marsh had "treated Mr. Harger and myself with entire fairness."

It was not spectacular wreckage. But Marsh and Powell should not have been relieved to read Ballou's pathetic attempt at big city journalism. Even a clumsy, boring attack in the *New York Herald* was a serious matter, and the dullness probably was further evidence of top-level involvement. The zombies wouldn't have dared to print such stuff without a go-ahead. Though Bennett wanted his paper to be read, he was quite willing to subordinate its quality to his whims. "I want you fellows to remember," the Commodore once told his executives, "that I am the only reader of this paper. I am the only one to be pleased. If I want it to be turned upside down, it must be turned upside down . . . If I say the feature is to be on black beetles, then black beetles it is going to be." Globetrotting Nellie Bly would be the toast of New York throughout the month, and the *Herald* was just beginning with its "hard nuts . . . for scientific digestion."

15

The *Herald* Steams Ahead

In a letter to Osborn, Cope sounded pleased with the January 12 issue: "You have seen Sunday's *Herald.* I have a good stock of shot and shell on hand as you see . . . There is nothing wrong with speaking the truth about Marsh, for his friends will not believe a word I say alone . . . I have, I believe, performed an unpleasant but necessary duty to my country, and I hope a clean job may be made of it." Even Wallace Stegner, an advocate not only of Powell but of good writing, grudgingly admitted his admiration. He wrote, "All the old charges were there in the *Herald*'s full and delighted story, all distilled and aged but not mellowed through twenty years of hatred . . . It was not, in spite of its hysterical extravagances, an attack to be laughed off."

It was one thing, though, for the *Herald* to brandish its trophies, another to impress the public. Judging from other New York papers, the January 12 feature caused little stir. The only paper to mention it the next day was the *New York Times,* which dismissed it in a short piece. A *Times* reporter had "thoroughly investigated" Ballou's "long rambling statement of charges" when he was "hawking them among the newspaper offices" on Christmas week. "Eminent representatives of geological and paleontological science," such as J. S. Newberry and James Hall, had assured him they were worthless. Professor Alexander Winchell of the University of Michigan had told him the controversy was old and deplorable and had been dismissed by the National Academy of Sciences. Cope had lost his fortune and was jealous of Marsh's wealth, and was angry at Powell for not hiring him. Professor Winchell considered Powell the man most suited to the post. The *Times* concluded that "the whole attack was declared to be inspired by a spirit unworthy of the persons engaged in it."

The only bit of wreckage to surface in the *Times* was a brief aside by Winchell. "I do not say that I approve altogether of what Professor Marsh has done, but such criticisms as I have to make I will make in his presence," he said. "I have told Professor Marsh that he ought to be more liberal with the younger men who were employed by him and ought to allow them the credit and honor of their discoveries."

The *Herald*'s article didn't play any better in the *Philadelphia Inquirer,* which ran a slightly longer Monday piece than did the *Times.* It interviewed Joseph Leidy, who attributed the affair "to Professor Cope's consuming restlessness," which got him into "hot water with his associates." Leidy praised Cope's work, and also Marsh's and Powell's, and he saw no grounds for the charges of plagiarism. "It is very unfortunate that this thing should have gotten into the newspapers," he concluded. "It will simply cause strife and vexation among the scientific men of the country." The *Inquirer* quoted Powell's counterattack at length and echoed Leidy in the conclusion of its piece: "The publication of the sensational scandal has caused the deepest regret among scientific men of the city." The article moved Cope to write to Osborn, "Poor old Leidy has come out against me . . . just as he has always done."

The *Herald,* undaunted by its rivals' neglect, devoted almost as much space to the feud on January 13 as it had the day before. Indeed, it deployed substantial resources to follow up its Sunday feature. "When hot news was involved, money was no object to Bennett," wrote the historian Richard Kluger. "No other paper spent so freely on the commodity basic to the trade." Monday's article included eight January 12 datelines from as far away as Canada and Ohio, indicating that many reporters and correspondents were telegraphing items. There was no editorial, but the headlines were strident enough:

VOLLEY FOR VOLLEY IN THE GREAT SCIENTIFIC WAR
Professor Cope Returns to the Attack, Ripping up Major Powell's Answer to the First Charges and Heaping More on Top of them

MORE LETTERS AGAINST PROFESSOR MARSH
Incompetency, Ignorance, Plagiarism Still the Main Offences in the Bitter Budget Charged Against Yale's Famous Geologist

MEN OF SCIENCE AGOG
Some Shocked, All Stirred Up, by the Sensational Disclosures in the Herald, and Many Unable to Believe the Accusations

LONG SMOLDERING EMBERS OF HATRED
Friends of Powell and Marsh Declare That The Pennsylvania Professor Is Inspired By Disappointed Ambition, Jealousy Envy and Unworthy Motives

"A majority of the scientific men of the United States were heard from, and their views were bitterly pronounced in not a few cases," the article began. "It is a very pretty fight as it stands." The first was Cope, in a story with Philadelphia in its dateline. The reporter complained of a search of several hours to find the "museum on Pine Street," making it clear that this was a genuine interview. Asked about Powell's counterattack, Cope dismissed it as "one of the stump speeches which is very useful in obtaining appropriations," and renewed his assault on Marsh. He decried the Professor's 1868 invasion of his Haddonfield Eden and accused him of smashing unwanted fossils, of buying fossils in Europe and renaming them in America, and of presenting other paleontologists' discoveries as his own.

"I should not have cared to expose Marsh's career had it not been for the injudicious zeal of his friends in elevating him to the highest place in the gift of American science," Edward concluded. "My patriotism rebels against this gross error and, for one, I wish to wash my hands of it. I refuse utterly to have my criticism of Professor Marsh put on the low ground of personal quarrel." Asked whether Marsh hadn't done "some good work," he replied, "Professor Marsh is a successful collector, and he has been very useful to science in making great collections which will be worked up at some future day. Had he been content to fill this role, the present exposure would have been unnecessary."

The Philadelphia reporter then consulted Persifor Frazer, but the copy he got was almost as boring as that of the day before. Asked about Powell's accusation that he and Cope wanted to take over the Geological Survey, Frazer called it "a presumptuous falsehood, for which I challenge Major Powell to present a scintilla of evidence"—then launched again into irrelevant geological protocols. The reporter hurried on to a more amusing chat with the University of Pennsylvania's Dr. Pepper, who denied that he'd threatened to fire Cope and maintained that he "had as a friend of the Professor endeavored to get him to abandon the project of making the charges against Director Powell and Professor Marsh." The *Herald* paraphrased Pepper with relish as he babbled on: "Now that the article had appeared he had no further control over the matter, but he supposed that Professor Cope, being a thinking man, would understand that he would be expected to substantiate the serious allegations he had made. In the event of his failure to substantiate them, and without any reference to his connection with the university, his value as a scientist would be, in a measure, impaired."

An item filed from Princeton showed equal pleasure in watching academe squirm. "Old Princeton has not had such food for succulent conversation in many months as this stupendous quarrel between scientists," it reported. "Of course, Princeton was too dignified to take any sides in the quarrel, but it could not help laughing up its sleeve over the neat and tasteful way in which Professor Cope had impaled Professor Marsh upon horns belligerent . . . For years the college authorities have chafed under the way Yale was getting all the geological plums at the expense of the rest of the collegiate world, and though they would not countenance the newspaper plan of exploding the scandal and thereby killing the evil, nor lend any assistance thereto, they were glad that it was done."

The Princeton reporter sympathetically, if disingenuously, devoted a column to W. B. Scott's complaints about Ballou, who had "put things in my mouth which I have never uttered and never thought." He was less indulgent of Henry Fairfield Osborn's airs and graces. "A very foxy professor is Professor Osborne," he purred, blandly misspelling his victim's surname. "He occupies the chair of biology at Princeton, and as proof that he is a learned doctor in the science of life, he can produce three of the handsomest children in the state of New Jersey . . . Professor Osborne looks like a college senior and an athlete, but he stands very high in learned circles, and has been named as a possible successor to Professor Marsh, when that gentleman gets out of the way. He put his hands in his trousers pockets and regarded me quizicaly [sic] when I asked him what he had to say about the Geological Survey and things.

" 'Yes, I have read the *Herald* article,' he said, 'with very great interest. I am glad the matter has at last come out. It will clear the atmosphere. The truth will be sifted out from falsehood and great good will be accomplished. I don't think I want to say another word.' " Rising to this challenge, the reporter probed for another word. " 'It has been stated,' said I, not knowing I was treading on Professor

Osborne's corns of ambition, 'that Professor Cope is the best equipped man in the country to fill Major Powell's position. Is that true?'

" 'Well, I don't know about that,' he said. 'Perhaps he is. Professor Cope is a scientist of high attainments and one of the most powerful and original thinkers alive.' Then, by way of variety, the handsome professor informed me that he had nothing further to say in the premises and I came away." The Princeton reporter also talked to Marsh's former assistant Otto Meyer, but found him less amusing. Meyer was evasive and denied that Marsh was sequestering the survey's fossils. He did say he'd left the Professor's employ because he was "disgusted" with his methods and that the Peabody lab assistants did most of the work, but that was old news.

A reporter who filed from Yale got little amusement from anybody; he telegraphed that the affair had caused "a great stir" but that nobody would talk about it, not even Samuel Williston. The faculty stood squarely, if mutely, behind Marsh. Professor William H. Brewer said "few would believe" Cope's charges, and particularly denied any suggestion of smashing or sequestering fossils. The disheartened reporter in New Haven subsided quickly, and those in other Ivy League towns fared no better. At Harvard, Professor N. S. Schaler, a Geological Survey member, "laughed heartily" when asked about the charges. At Cornell, Professor H. S. Williams simply called them "not fit to publish."

The *Herald*'s Washington correspondent began more hopefully. "It is rare that the Kosmos [sic] Club has been so excited of a Sunday as it has been today," he wrote. "This club is the headquarters of the many scientific men who reside in Washington, and as soon as they had read their *Heralds* at their breakfasts at their homes they hurried to the Kosmos. There all the points in the *Herald* article . . . were gone over and eagerly discussed by knowing hands." The correspondent actually found a crack in the establishment wall when C. Hart Merriam, chief of the Agriculture Department's Division of Ornithology and Mammalogy, observed "that there is a good deal to be said on both sides when you come to the personal quarrel between Professor Cope and Professor Marsh." Merriam also said, however, that Powell had made the survey "the grandest scientific institution the world has ever known," and that Cope was a jealous and disappointed man. Clarence E. Dutton, chief engineer of the Geological Survey, agreed, not surprisingly. Professor Thomas Wilson of the National Museum said he was friendly with Cope and Powell but had never met Marsh and knew nothing of the matter. Representative Hamilton Dudley Coleman of Louisiana also "knew nothing about the matter, and consequently had nothing to say," but promised that if it came before Congress, "he would endeavor to master it, and the *Herald* might be confident that he would champion the side of justice."

The *Herald* had to leave the eastern seaboard to get any criticism of the U.S. Geological Survey, and even what it found was muted. In Marietta, Ohio, Cope's old friend from the Secretary of Interior's waiting room, General Warner, sought out a *Herald* correspondent to say that, as a congressman, he had looked into the survey and had found considerable favoritism and busy work "of not much scientific value." In Ottawa, Assistant Director Robert Bell of the Canadian Geological Survey refused to comment on the Cope-Marsh rivalry, but admitted that he considered Powell's survey a parvenu, slovenly affair in proportion to the "lavish" government appropriations it enjoyed. "The leading fault is that they are too ready to rush into theories based on a meagre amount of facts. It is not necessary for me to say that is not the scientific spirit of today. Facts, facts, facts!"

The January 13 coverage ended on that *diminuendo* note. Despite Cope's initial eloquence and Bell's and Warner's barbs, it wasn't a good return on Bennett's dollar. Even Cope sounded downcast in a letter he wrote to Osborn that day. "There is one favor I ask of you and of Scott, that is if opportunity occurs to state whether you have found me so 'Jealous and suspicious that it is impossible to cooperate with me.' I may be wrong but I do not think I am very jealous! . . . Scott, yourself, Leidy, and I are the only persons acquainted with Marsh. It will now rest largely with you whether I am supposed to be a liar and am actuated by jealousy and disappointment."

The *Herald*'s January 13 feature certainly didn't impress the New York competition. The *Times* had sent a reporter to ask Marsh whether he and Powell "had combined to enrich themselves in repute and pocket at the expense of the government," but seemed content with his response that the charges were absolutely false and his accuser was "crazy or made crazy by jealousy." It interred the matter in two paragraphs on page six of its January 14 edition. The *World* and *Frank Leslie's Illustrated Newspaper* were too full of "The Baby King of Spain," "The Saint Louis Cyclone," and of course "Globetrotting Nellie Bly" to accommodate academic quarrels, although the *World* did mention in its "personals" column that "the students of Professor Honey at Yale are not sweet on him." The *Tribune* devoted an entire article to Professor Honey's troubles, but made no reference to the paleontologists' fight except to reprint a few paragraphs from the *New Haven Palladium* of January 13.

"Professor Marsh of Yale may dismiss the sensational charges of Professor Cope," the *Palladium* had declared, "with the consciousness that his antagonist is not one of sufficient repute as a wise man and of good judgment to harm the sage of Yale . . . The truth seems to be that Cope is a disappointed scientist of a good deal of ability. He is represented as a visionary, hot-headed man who has had the ill-luck to lose a fortune in chasing the country over in search of elephants' backbones and the ear-marks of animal antiquity generally on the shifting wastes of time . . . Altogether, Cope is to be pitied. He should have been content quietly to peck away at the curious conformation of beautiful Schuylkill's banks."

Bennett wasn't finished throwing money at the problem, however. On Tuesday, January 14, the *Herald* returned to the attack with both a feature and an editorial. "The Cope-Powell-Marsh controversy which has been going on with unusual vigor in the *Herald* has very naturally created a sensation in the scientific world," the editorial boasted, "and very naturally it has divided the distinguished scientists who have spoken on the subject." Then it got a little defensive. "The *Herald* is simply the forum in which the speakers of both sides are heard," it assured its readers. "If the parties were waging a merely personal quarrel it would be an affair of their own. But on one side the representatives are government officials, and the issue raised by a scientist of recognized standing involves an important breach of the public service." Shorter but still strident headlines introduced the feature:

WIDENING THE GEOLOGICAL CHASM
Professor Cope and Director Powell Still
Engaged in the Pleasant Pastime of
Damaging Each Other's Sci-
entific Reputation

GIVE AND TAKE SEEMS THE RULE
Like the Kilkenny Cats, If the Squabble Much
Longer Continues There Won't Be Much
Left of Either Combatant

ROOM FOR PROFESSOR MARSH

Carelessly casting aside the editorial's avowed neutrality, the article began by trying to patch Cope's tattered reputation with scholarly European testimonials. Professor L. Rutimeyer of the University of Basle likened his work to Cuvier's. "The scientific public impatiently awaits the completion of this enterprise, which is considered by every one to be worthy of the great country from which it issued." Professor Albert Gaudry of the Jardin des Plantes in Paris expressed like sentiments, as did Professor Karl von Zittel of the University of Munich, who predicted that the United States would "add a previous leaf to its wreath of fame" by publishing the second volume of Cope's "Bible."

Osborn and Scott chimed in, and the *Herald* even brandished some pro-Cope copy from another newspaper, the *Philadelphia Record* of January 13. "At last the scientists of this country have entered upon a campaign against Major John W. Powell and the Geological Survey Bureau," the *Record* had exulted. "In the opinion of Professor Cope . . . Major Powell knows a great deal more about politics than he does about geology, and in regard to the gallant Major's scientific attainments no man is more competent to speak than the learned paleontologist of the university."

Ignoring the resolutely anti-Cope papers, the *Herald* sank deeper into journalistic favoritism with a full-fledged sob story headed ALL FOR SCIENCE. To the throb of violins, as it were, a zombie told how Cope had spent almost every penny of his large fortune on scientific research and now depended on his meager salary while Marsh and Powell, respectively wealthy and well-to-do, were "so well intrenched and buttressed" in the survey that it seemed almost quixotic for Cope to try to dislodge them. "Nevertheless, although his adversaries are rich, powerful, and have influential friends in almost every department of the government," the zombie declaimed, "the plucky Pennsylvania Professor springs as gallantly to the fray as any knight of old."

The *Herald,* though, wasn't betting as much on the plucky Philadelphian as it had the day before. It ran only four datelined items, and they were brief. POWELL AGAIN MAKES DENIAL began the first, from the Washington correspondent, who clearly was fascinated with scientists *en masse.* "There was a procession of scientists marching in and out of the *Herald* Bureau all day," he wrote. "They came to ask for copies of the *Herald* of yesterday and today . . . Some said the charges were scandalous and outrageous; others that Professor Cope was instigated by jealousy and was mad because he did not have as good a pull on the treasury surplus as those whom he attacked. Some regretted the exposure while others thought it was a good thing."

The correspondent tore himself away from the procession long enough to visit Powell's office and demand "if he had carefully read the *Herald* of yesterday and today." The major met this impertinence with his usual good grace, and was "glad to reply," when asked whether he'd unduly favored congressmen by hiring their sons, by stoutly denying it. Powell also "rapped Dr. Bell's knuckles" for the Canadian geologist's slurs against the U.S. Survey. After this friendly exchange, the correspondent hurried back to his scientists, noting that "the annual meeting of the Cosmos Club, occurring this evening, brought together a larger number of members than usual," and then gave his assignment a final nod. "So far as the war inaugurated by Professor Cope was referred to," he concluded, "the general drift of talk seemed to be in favor of Major Powell."

The other items were even briefer. The Philadelphia reporter quoted Cope's response to Professors Shaler's and Williams's charges of eccentricity and jealousy. Edward called Williams a "blatherskite," and said his only eccentricity lay in "his unwillingness to stand by and see imposition," adding that he'd bring up more "documents and sources of information" after Marsh's reply, although Marsh probably would "make no attempt at defense." A New Haven item had Marsh promising that his reply would appear "as soon as possible," however, and Williston promising to reply to Marsh's reply. "The prevailing opinion in this city seems to be that Professor Marsh's reply will be quite exhaustive," the item concluded hopefully.

The last item, from Cambridge, suggested that the *Herald*'s attention was wandering. It was an interview with "Assistant Samuel Gorman at the Museum of Comparative Zoology," who said he'd worked under both Powell and Cope. "I know that they are both very able men," he said, "and I cannot imagine that either of them would be guilty of intentional plagiarism or any other breach of gentlemanly conduct." Since Marsh was the one accused of plagiarism, this odd statement may have been a misquote. But the *Herald* had netted an odd fish. "Gorman" was Samuel *Garman,* the self-styled Quaker schoolteacher whom Cope had fired for extortionate pay demands at Fort Bridger in 1872.

"Apparently deeply affected by the Cope-Marsh feud," said the biography of Garman in the papers of the Society for the Study of Reptiles and Amphibians. "He was very secretive about his research, refusing to talk about it or to show specimens even to most of his museum colleagues. He worked in a dimly-lighted quarters of the museum, and when someone knocked on his door, he would cover all specimens on which he happened to be working before going to the door, and even then cracked it open only enough to talk with the visitor." An alert reporter would have sensed opportunity in such behavior, and Garman, if questioned adroitly, might have said interesting things. The *Herald* let him wriggle away, muddying the waters as he went.

From his "cracked-open" door, the museum assistant said he "knew nothing at all about the truthfulness of the statements made by the Professor." When asked whether Marsh had smashed fossils, he "said that he did not think that the leader of an expedition would do any such thing," and then indulged in deliberate obfuscation in the guise of scholarly confusion. As though the *Herald* wanted to know whether smashing fossils was standard scientific practice, Garman patiently explained, "The usual method followed by the expeditions with which I have been connected was that, when a large number of fossils had been collected we would bury all that we could not take away and mark the spot in such a manner that we could direct others where to find them in case we wanted them later."

16

Marsh Strikes Back

The great scientific war had been a dud so far. Even Cope seemed disappointed, and he evidently hadn't learned much about newspapers, still thinking his hack ally was in control. "Ballou misrepresented me," he wrote to Osborn on January 14, but added, "I am glad he started the battle just as you and Scott say you are, and am not going to abuse him for his spread eagleism, although he has hurt me." Cope was smarting that morning from another *Philadelphia Inquirer* article headed COPE MAY BE REMOVED. The *Inquirer* reported a "well-grounded rumor" that the University of Pennsylvania trustees would ask for Cope's resignation unless he could substantiate his accusations against Marsh and Powell, although an interview with the beleaguered provost did not suggest any great institutional resolve. "When asked about rumors," the *Inquirer* reporter wrote, "Dr. Pepper said: 'I am not acquainted with the subject.'

" 'But don't you know that the *Herald* published nine columns on the affair last Sunday?' he was asked.

" 'The probably accounts for why I know nothing of it,' he replied quizzically."

Despite Dr. Pepper's spot of oil slick, the *Herald*'s wreckage collection was so meager in proportion to its eager anticipation of government scandal and shake-up that the matter probably would have ended there had Marsh kept silent. His delay in answering the charges suggests that he'd been weighing the options. Like Cope, however, he clearly couldn't resist venting his spleen. He'd probably been compiling "Copeiana" for as long as Cope had been filling his Marshiana drawer.

The zombies must have been relieved when they finally got Marsh's reply. Comparing it with Powell's, Wallace Stegner observed that it was "marked by no such restraint and decorum. Having been attacked with talons, he replied with claws." If Powell had tried to torpedo the *Herald,* Marsh surfaced in full view, his cannon blazing. It was what the zombies needed to keep the story going, and they gave it "full and delighted" treatment on Sunday, January 19.

An editorial headed PROFESSOR MARSH MEETS HIS FOE belabored the Commodore's martial metaphors, calling Marsh's counterattack "entirely unique . . . as a specimen of the rhetoric which an indignant scholar can use when occasion requires it," and observing that the Yale Professor proposed "to use whatever weapons the God of nature has placed in his hands." The writer may have had in mind a visit to Slaughter's Gap. "It is but fair to say that the two contestants have not fought shy of each other," he concluded. "They have met on the field with the determination that one of them shall be carried off in an ambulance to the nearest coroner . . . We shall leave to the scientists the sad duty of looking after the wounded or burying the dead." The feature's headlines were as lengthy and strident as the previous week's:

MARSH HURLS AZOIC FACTS AT COPE
Yale Professor Picks up the Gauntlet of the Pennsylvania Paleontologist and Does Royal Battle in Defense of his Scientific Reputation

HURTFUL ALLEGATIONS DENIED

Deft parrying of clever thrusts at the Administration of the Geological Survey and the National Academy of Sciences

WAR CARRIED INTO AFRICA

Director Powell sustained in Most Vigorous Fashion and the Assailants of His Governmental Bureau Put in a Defensive Attitude

EPISTOLARY CORROBORATION

Perhaps Some of the Scientists Concerned in the Geologic Strife Who Have Heretofore Kept Silent Will Now Make Explanation

NOW IS YOUR CHANCE, GENTLEMEN!

The headlines belied a wilting of editorial resolution, however. Any prospect of government scandal had faded, and even scientifically inclined readers who'd soldiered through the previous week's paleontological confusions may have been put off by the growing flippancy manifested on January 14 in the reference to "Kilkenny cats." The scientifically meaningless hurling of AZOIC FACTS would have irritated naturalists, while THE WAR CARRIED INTO AFRICA would have puzzled them. Would the professors fight over African fossils now? Allied to the levity, undoubtedly contributing to it, was a note of desperation that sounded clearly in the last headline. Now, in fact, was the *Herald*'s last chance to goad into action an aloof scientific establishment.

Irresolution showed even more clearly in the article's opening paragraphs. The zombies blew another fanfare, but it was a tired one this time: "charges . . . alleging the grossest ignorance, incompetence, and plagiarism . . . important government departments entrusted to their care . . . tremendous sensation among the scientists . . . echoes of the explosion have not yet ceased to be heard." Then they abruptly unleashed Marsh.

The Professor didn't waste time graciously thanking anybody, least of all the press; he went for blood. Cope was a slanderer who had "devoted some of his best years" to attacking America's premier paleontologist, "and it may thus be regarded as the crowning work of his life." Marsh had "suffered those attacks in silence" for years "because it seemed to be due to the positions I have held to abstain from all personal controversy." Now controversy was forced on him. "To meet these charges one and all is an easy task, but not a pleasant one, as I shall have to use plain words and say many things which I should otherwise wish to leave unsaid." Powell's case against Cope had been a Methodist one; he had painted him as a sinner deep in the depravities of jealousy, hysteria, and "vicious species work," though still redeemable if he saw the error of his ways. Marsh's was Calvinist. Cope was damned, and Marsh consigned him to perdition as unyieldingly as a Massachusetts ancestor might have condemned an unrepentant Quaker.

A brief lull followed Marsh's initial salvo. Some of the *Herald*'s depth charges had to be dodged before he could get on with scuttling Cope. The most accurate was that he had colluded with Powell to control the National Academy of Science's committee on survey reorganization in 1878, but the Professor evaded it with some rhetoric and name-dropping. The academy's honor was as sacred to him as his own, and Cope had "chosen to tarnish its fair name." Only the most upright and impartial academy members had served on the committee, and Marsh listed them with a reasonable assurance that *Herald* readers wouldn't know they were his cronies. Anyway, he concluded, the academy had tried several times to expel Cope because of "charges of such a serious nature that they would prevent his holding any position in any institution of higher learning . . . Need I say more in reply to charge number one?"

Marsh evidently felt safer with charge number two, because he went for blood again. Did Cope accuse him of mixing the government's fossils with Yale's and denying the public access to them? Perhaps he did restrict access a bit, but that was because there were "scientific swindlers" lurking in the halls, as in the case of Cope's 1882 visit to the Peabody laboratory. A liar and slanderer, Cope also was a thief, barefaced and without remorse. "Had Professor Cope been a man of honor, he would have been humiliated by what he had done and made prompt reparation," he wrote. "On the contrary, he boasted of his act and has since continued to publish the result of what he saw with many falsehoods added . . . This raid by Professor Cope on my museum is only one of many similar acts in which nearly the whole scientific world has suffered."

In a one-man crime wave, Cope had plundered not only the Peabody Museum, but British, Berlin, and Paris museums, the Harvard Museum of Comparative Zoology, the Peabody Academy, the Columbia University collections, and "especially" the Philadelphia Academy of Natural Sciences. "I have the facts from the authorities of each and every one of these museums," Marsh clamored, "and very damaging facts they are." Cope also routinely raided private collections, most notably Marsh's, as revealed by his 1873 letter accompanying the Kansas fossils that "had been abstracted from one of your boxes." Othniel judged that "the Sunday *Herald* with its supplement would not contain the half" of his evidence of Cope's "depredations."

Leaving more blood in the water, the Professor then glided swiftly past Cope's charges of his misusing funds, destroying fossils, and underpaying, neglecting, or otherwise maltreating his assistants. Again, rhetoric and name-dropping sped his way. Buffalo Bill was invoked, as were W. N. Rice, H. S. Williams, George Bird Grinnell, and other early assistants who had gone on to attain scientific distinction. Marsh acknowledged Oscar Harger's help, although "during the last half of his service he was an invalid, face to face with death, and I endeavored to repay, by kind attention and ever watchful care, the services he had rendered in health."

As to the other assistants, they "were nearly all employed in mechanical or clerical work alone, as most of them were not sufficiently versed in scientific work to make their services of special value to me." Flattered and suborned by Cope, "men whom I had employed simply to clean fossils or measure them became at once profound anatomists . . . A touch of Professor Cope's magic wand, and again the same men became authors, who kindly wrote the works that I had for years in preparation. As the spirit of his own vanity pervaded them, they closely imitated their master, and proved his apt pupils in prodigality of accusation and economy of truth. Their ranks were joined by a few other young partisans of this disappointed leader, and we have their combined efforts in the present attack."

The enigmatic African headline came into play at this point. "Little men with big heads, unscrupulous in warfare, are not confined to Africa," Marsh continued, "and Stanley will recognize them when he returns to America. Of such dwarfs we have unfortunately a few in science, and some of them have fallen ready victims to the wiles of Professor Cope's flattery and promises of friendship. How reliable his friendship is many, both dwarfs and larger men, have learned to their cost." Stanley was news again in 1890 because of his third African expedition—to "rescue" the naturalist-explorer Emin Pasha, another of Conrad's *Heart of Darkness* models—so Marsh's reference simply may have been topical. Yet it was such a non sequitur that it may have been more deliberate. Othniel's celebrity career would have made him much less naïve than Cope about newspapers. He probably knew how absolutely Bennett controlled the *Herald,* and how much the dissolute publisher hated Stanley, so his reference could have been a veiled taunt at Bennett the "dwarf." In turn, the WAR CARRIED INTO AFRICA headline might have been an ironic *touché* exploited as a come-on to jaded readers.

Having disposed of Cope's political charges, Marsh took up his accusations of plagiarism, the most serious ones from a scientific viewpoint. "A slander more false and malicious has never been made against a man of science," he proclaimed, and produced a convincing testimonial from his former assistant George Bird Grinnell that he had dictated or handwritten most of his toothed-birds monograph, less convincing statements from Williston, Baur, and Schlosser that he was the author of his *Dinocerata.* He insisted that Oscar Harger had been too sick to have written the monograph, and the deceased Harger did not contradict him. He then turned his fire on Cope's initial charge, that he had plagiarized the Russian paleontologist Kowalevsky's work on horse evolution. It had been an unwise one, because Marsh's horse work was his strongest. Briefly lowering the volume, the Professor modestly related how he had convinced the great Huxley of his own conclusions as to the American genealogy of horses. Then he turned it up again and used Kowalevsky, whose career had been checkered and tragic, as a human pitchfork to shove Cope deep, deep into the eternal flames.

"During the recent International Geological Congress in London," Marsh declared, "I attended a conference of museum authorities who met to discuss the depredations they had suffered from visiting scientists. The cases of Cope and Kowalevsky were fully discussed and the extent and skill of their respective work were topics of lively interest. The general impression was that it was a close race between them and no competitors in sight. Like the famous race between the dog and the wolf it was a case of nip and tuck, and the general impression seemed to be that, as in many more honorable international contests, the American was a little ahead. Kowalevsky was at least stricken with remorse and ended his unfortunate career by blowing out his own brains. Cope still lives, unrepentant."

Marsh might well have ended on this hellfire note. His indictment of Cope as a museum bandit was striking, and would have been damning had he cited evidence. Even without it, the attack would have resonated if Marsh had closed with a brisk Calvinist version of Powell's *ad hominem* summation. But the Professor couldn't resist a chance to "spread himself," and he kept firing salvo after salvo of self-justification. He reviewed the technicalities of Cretaceous mammals, horned saurians, Mesozoic birds, Permian reptiles, and Kansas pterodactyls. He deplored Cope's ownership of *The American Naturalist.* He told how he'd met the unbalanced Edward in Berlin, and how the younger man had turned against him because he'd revealed his misplacement of the plesiosaur's skull. Finally, he made his worst journalistic mistake, one worthy of Ballou. He presented the *Herald* with the eleven-page attack on Cope's uintathere studies that he had published in *The American Naturalist*'s bad-boy appendix of June 1873. The newspaper public would have found it, even intact, wholly unreadable, and the *Herald* zombies made it incomprehensible by chopping it into a few paragraphs.

Marsh evidently sensed he'd gone on a bit, because he tried to liven things up at the end. He quoted from Cope's lame July 1873 *American Naturalist* appendix response to his lengthy June attack. "The recklessness of assertion, the erroneousness of statement and the incapacity of comprehending our relative positions on the part of Professor Marsh render further discussion of the trivial matters on which we disagree unnecessary etc." Then he concluded: "Professor Cope's feeble response shows plainly that he was then in the exact condition of the boy who twisted the mule's tail. He was not as good-looking as he was, but he knew more. Has Professor Cope since learned wisdom from his increasing years? The public must judge. The scientific world has long since passed judgment."

Marsh must have taken almost as much satisfaction as Cope had in publishing his docket. If the *Herald* had expended its testimonials and sob stories in support of a liar and fossil thief, there had to be some wreckage on the Commodore's deck. Even Henry Fairfield Osborn had to admire "the thoroughness and ability" of the attack. "Whereas Cope attacked after a truly Celtic fashion, hitting out blindly and with little or no precaution for guarding the rear," he wrote, "Marsh's reply was thoroughly of a cold-blooded Teutonic, or Nordic, type, very dignified, and, under the cover of wounded feelings reluctantly breaking the silence of years, as if this reply had been forced on him."

Yet the Professor's barrage probably ended with even fewer readers than Ballou's screed. Fossil theft did not tickle the 1890 public's palate for mayhem any more than did scientific plagiarism, and once the mystery of the African dwarfs had been solved, even amateur naturalists must have drifted away into the Sunday morning air. If Othniel had anticipated admiring journalistic comment on the majesty of his wrath, moreover, he would have been disappointed. The *Times* and *Tribune* paid even less attention to the *Herald* of January 19 than they had to that of the previous Sunday. The *Inquirer* also ignored it. The response was so meager that Marsh was reduced to sending copies to his hirelings. "it must have made Cope feal ashaimd of him Self," a Nebraska bone sharp named Gus Craven wrote back dutifully, "but it may be as you say their is not mutch left in him."

17

The *Herald* Steams Away

Marsh was reasonably accurate when he said the scientific world had "long since passed judgment" on Cope. It hadn't jailed him as a museum bandit, but that might not have been much worse than years of exclusion, for someone as proud and ambitious as Edward. The matter of the *public*'s passing judgment on Cope, or on Marsh and Powell, was more of a problem. How could the public judge an affair that largely was beyond its comprehension and that the *Herald* had further obfuscated with its interminable, chaotic coverage? That coverage so dazed Wallace Stegner that he thought Cope's feeble reply to Marsh's attack in the June 1873 *American Naturalist* —which Marsh quoted in the *Herald* seventeen years later—was his response to Othniel's 1890 barrage.

"Cope had shot off all his ammunition in the first charge," Stegner wrote. "He was scattered and routed by Powell's dignified immovability and by the bullwhip of Marsh's tongue. Given opportunity to make fresh statements so that the *Herald* could keep its profitable controversy going, he replied only that 'the recklessness of assertion [etc.]' . . . But he was not to get away with such a lame and lofty curtain line. Marsh would have the last word." Stegner then paraphrased Marsh's taunt about the mule's tail as that last word.

The *Herald*'s feud coverage did not end with Marsh's counterstrike, however. The paper sent a reporter to interview Cope about it (probably the same reporter who had gone the previous week, since he didn't get lost) and ran the interview on page 3 of its Monday edition, January 20. Although there was no editorial, the headlines remained martial:

SCIENTIST FIRES BACK AT MARSH
He Denies the Charges of the Yale
Professor, Admits Some Minor Errors and Pours Fresh Shot
Into the Enemy

FORMER CHARGES REITERATED
In the Opinion of the Plucky Pennsylvanian
The Paleontologist of the Geological
Survey is not Competent to Criticize Fellow Scientists

DR. BAUER [SIC] INDIGNANTLY RESIGNS
He Declares that Professor Marsh Attempted
to Mislead by quoting only a Fragmentary sentence in his letter

PROFESSOR WILLISTON EXPLAINS

PROFESSOR COPE ON THE WARPATH a subhead declared, and Cope energetically set out to reinforce his charges against Marsh. "I have another broadside for Marsh as soon as his reply comes out, a new catalogue of his robberies since 1873," he'd written to Osborn the week before. "It is equally forceful with anything that has appeared." He evidently did not feel that any conclusive judgment had been reached.

But something had changed. After he'd generated a few paragraphs of anti-Marsh rhetoric, including his view of the stories about their early acquaintanceship and the head-to-tail plesiosaur,

Cope suddenly found himself the subject of an interrogation about Marsh's accusations. Had he attacked Marsh and Powell because of malice or insanity? Had he stolen Marsh's ideas and fossils? Had he inaccurately dated publications? Was he jealous and disappointed, opposing Marsh "for the mere sake of being an obstacle to him, a thorn in his side, so to speak"? Cope did a creditable job of parrying these pointed questions. The reporter, though, displayed such brisk journalistic impartiality that the "plucky Pennsylvanian" may have wondered whether the man was from the same paper that five days earlier had sobbed about his having given "all for science."

In fact, the martial headlines were an ink cloud. Whether or not Marsh was a plagiarist or Cope a thief, "the gentlemen of the scientific world" clearly had missed their chance to fight over the matter for the *Herald*'s benefit. "Whenever there is an important piece of news, I want the *Herald* to have the fullest and best account of it," Bennett once had written to his managing editor. "Another point which I think you understand is letting a thing drop the moment public interest in it begins to flag. The instant you see a sensation is dead, drop it and start in on something new." This cynical advice, of course, had been less than candid, since the Commodore didn't mind featuring something that bored or mystified the public as long as he was interested in it. But Bennett's interest in paleontologists must surely have evaporated after Marsh and Powell slipped the net of Red Cloud–style scandal. The hard-driven *Herald* zombies had to move on, leaving whatever wreckage might bob in their wake, and they did so remorselessly—as Cope discovered in his final interview.

The only real news in the January 20 article was George Baur's resignation, and the story added little to its misspelled headline. Baur had resigned out of pique at Marsh's highhanded use of his name to deny the plagiarism charge, but the article covered it so vaguely that the resignation caused no ripples. The paper got Williston's first name wrong, too, and rendered his pique even more vaguely, indeed, incomprehensibly. "W. Williston was also an interested reader of the reply," the reporter rambled, "and in view of the feelings that existed between those two learned scientists the statement given me today by him is of decided interest to Yale and her friend."

"It is evident that the reply of Professor Marsh is not the conclusion of the whole matter and that some pretty interesting news will be gleaned from the subject in the near future," the *Herald* blustered, but the next item gave away what it really thought. The levity that had crept into the January 14 "Kilkenny cats" headline prevailed, and the January 20 coverage ended with some of the "comic verse" that Victorians produced so abundantly when they could think of nothing else to do. It printed the four stanzas of "Paleozoic Poetry: The Unfortunate Pterodactyl Wings Its Flight Through Prosody," signed by a forgotten figure named Albert Edmund Lancaster. The first stanza read:

Professor Cope to Professor Marsh:—
Your ignorance of saurians is something very strange:
The manners of the Laramie are far beyond your range,
You fail to see that certain birds enjoyed the use of teeth,
That pterodactyls perched on trees, nor feared the ground beneath.
You stole your evoluted horses from Kowalevsky's brain,
And previous people's fossils smashed, from Mexico to Maine.
To Permian reptiles you are blind, in short, I do insist;
You are—*hinc illae lachrymae**—you are a plagiarist!

In the guise of good-natured fun, the poem expressed some sourness at the scientific flash-in-the-pan on which the *Herald* had spent so many man-hours. It showed what the zombies really thought of the savants whose egos they'd been stroking, however insincerely, for three weeks. Its "Moral" was:

* hence these tears

So Science walks, with gait serene, her crown an olive sprig,
Intent alone on holy truth and *otium cum dig**.

The acid bit deeper when the *Herald* briefly returned to the affair on Sunday, January 26. It published a letter from the former Peabody assistant Otto Meyer about Marsh's familiar laziness, ignorance, incompetence, and dishonesty, but that was merely a pretext for an editorial, entitled THE BATTLE OF SCIENTIFIC GIANTS STILL BEING WAGED, which lampooned the paleontologists cruelly. "Professor Cope recently threw his red hot shot onto the camp of Major Powell," it announced. "All of the world was at once agog. Cope is not a trifler. He meant business, and there was blood on the face of the moon. He made charges against Powell and Marsh which can never be forgiven—that the one was an incompetent and the other a plagiarist. Not even a paleontologist can condone an offense of that kind, and in our mind's eye we saw the contestants, like Shylock in the play, sharpening their knives for their pound of flesh. Marsh replied in our columns with vigor, indignation, and an evident determination to attend the funeral of Cope. He metaphorically swore at his accuser in all the dead languages at once, and flung scientific terms at his head like so many cobblestones. The spectacle was worth viewing, from a safe distance. Then Cope trained his guns on Marsh once more, and we expected a sudden call for an ambulance. His detonating rhetoric was like the echo of distant thunder, and the heavens looked pretty blue."

It concluded: "And this morning, Dr. Otto Meyer . . . strides into the field, armed cap-à-pie, and draws a sword as long as that of Richard the Lion Heart. He smites with the muscle of an athlete and without mercy. Well, these little disagreements serve to break the monotony of life, and our readers will listen to the story of Dr. Meyer with respectful and interested attention." They would have if they could have found it. The *Herald* showed its estimation of

* idleness with honor

Meyer's letter by running it on page 25, beside a piece labeled MISSION WORK AMONG THE COLORED PEOPLE. The letter was subtitled BIG MEN WITH LITTLE HEADS.

The *Herald* dropped the paleontologists after that, and the silence of the other New York papers was deafening. Only the *Tribune* even skirted the subject, with a January 27 editorial that pointedly ignored the previous day's lampoons in the *Herald.* "Topographical surveys intended to show how far it is possible to reclaim the arid belt of the West have been going on under Major Powell's able direction for some time . . . ," it said. "Farming by irrigation is certainly the most scientific way."

The sudden silence makes it hard to surmise what people thought in the affair's immediate aftermath. William Hosea Ballou again demonstrated his unreliability when he wrote, in his 1908 *Chautauquan* article, that his "fifty-two-column exposé" had "smashed" completely "one of the worst rings that ever fastened on the government." His account of ensuing events was even more self-aggrandizing. "Cope's works were all printed soon thereafter and he received all his back pay. I have forgotten what Cope said and wrote to me about the matter, something pleasant I think." None of this happened because of Ballou's "exposé," but what did happen has been less clear.

Wallace Stegner's confusion of Cope's 1873 reply to Marsh with his 1890 one has fostered an impression that Marsh got the better in the *Herald* fight. Stegner thought that Cope had been "discomfited and even discredited in the eyes of most scientific men," and Robert Plate later wrote that "Marsh's able rejoinder against Cope's reckless assault entitled him to a partial victory." Certainly, Marsh's social and professional status suffered no immediate damage after the *Herald* affair, unless the establishment's embarrassed ignoring of his vicious counterstrike could count as such. A few months later, Andrew Carnegie wrote him a glowing letter, referring to plans to put some of his steel-making millions into museums. "I astonished the committee with the *Brontosaurus excelsus*—knocked them flat—every one," the tycoon wrote. "Told them the first thing that went into the museum had to be *that*—the first contribution made by the first collector of the world!" It may have seemed at the time that the *Herald* articles had hurt only Cope.

The recollections of two feud participants suggest that the situation was more complex, however. In *Master Naturalist*, Henry Fairfield Osborn recalled that "the Cope-Ballou article proved to be a boomerang" because of Marsh's counterattack, but concluded, "On the whole, it is well that this great explosion took place." W. B. Scott thought Marsh had been "unassailable" because of his "strong following, especially in New England, of men in other branches of science . . . a position of great influence, from which newspaper squibs were not likely to dislodge him." Scott did recall, though, that the Peabody assistants' statements against the Professor had been "very damaging," and he didn't even mention Marsh's counterattack. Osborn's and Scott's remarks imply, at least, that Marsh would have been better off had he not counterattacked in the *Herald*, since it did his "unassailable" position no good and prolonged the affair.

Actually, Marsh was less unassailable than he may have seemed. In the process of boring its readers, the *Herald* ran articles that smeared him and Powell pretty widely. Even the *New York Times* and *Philadelphia Inquirer* spread some mud simply by pooh-poohing the feud, and other newspapers were less dismissive. The *Washington Star* devoted four columns to the story, and the influential *Chicago Tribune* ran a tart editorial.

"A scandal of no ordinary dimensions has broken out in the Geological Survey . . . ," the Chicago comment began. "The arena of the controversy is daily widening, and it is not impossible that all the scientists in the country may yet take a hand in the bitter fight." The editorial gave a short but perceptive account of the issues involved—perhaps the clearest and most impartial newspaper account—before it too descended into *Herald*-style burlesque. "All savantdom is agog and ready for the fray," the *Chicago Tribune* declaimed, freely appropriating the Commodore's bombast, "and any day may witness a general onset and crash between the consolidated columns of geology and paleontology, with all the other ologies careering about as sharpshooters and outriding guerrillas." Such medleys of mockery and substantive-sounding slurs could be as damaging in the long run as a congressional investigation.

In an editorial he wrote for the February 1890 *American Naturalist,* Cope hardly sounded discomfited. He thumbed his nose at the scientific establishment: "The press has taken hold of a question of vital interest to the science of this country, which too many of the scientific men themselves have been unwilling to touch. It is unfortunate for the reputation of some of our scientific men that they have neglected the matter so long that its adjudication has been passed into the hands of the public. The matter should have been quietly disposed of among themselves, but it has now gone before a wider tribunal in which the susceptibility of individuals will be less considered." Insisting that his foes had been serious about grabbing his fossils, he further accused Powell of refusing to publish his second "Bible" volume because it would have "anticipated" Marsh's work: "The man who hired others to do this work could not tolerate another man who did his own work so 'near the throne.' Besides, he could not do the work without the specimens used by his predecessor, the other man, and so he must get possession of them, although the private property of the latter."

Cope likened himself to Naboth, who was slain so that the evil Old Testament King Ahab could take his vineyard. "The modern Naboth, however, lived in the land of newspapers and of public opinion," he wrote, "and these have been heard from. Ahab has not yet obtained the vineyard." The final sentence showed that he still felt threatened. "I have been in Washington and the Sec. Interior told me that he was satisfied that my collection belonged to the U.S. Govt!" he wrote to Persifor Frazer in March. "This huge steal appears to have taken hold in Washington . . . I will busy myself in exposing this villainy. Fortunately, I have some strong ammunition." For a while, Edward kept slinging defensive mud; in the March and April *American Naturalist,* he published Baur's and Barbour's slurs on Marsh's paleontological methods. Yet he evidently felt more justified than discredited by the *Herald* affair.

18

Symmetries and Ironies

Cope stopped his active pursuit of Marsh and Powell after 1890. To a degree, the bitter medicine of the *Herald* affair seems to have cured his Marsh fever. Being called a jealous, demented thief in the newspaper was a shock, as his letter of January 14 to Osborn showed, and perhaps this lanced the carbuncle of inflamed vanity that had tortured him since Fort Bridger in 1872. At least, finally venting the anger he'd brooded over for two decades must have been a relief. Now there was an air to Cope's life of a man freed from a compulsion. He still complained of Marsh's behavior, as in a letter to Persifor Frazer in 1892 about the Professor "making an exhibition of himself again. He has been redescribing and rediscovering most of my important discoveries in paleontology of mammalia for the last 10 years!" But he calmed down.

Cope had another good reason to stop pursuing his adversaries. Abler hands soon picked up the ball that he and Ballou had fumbled onto the field. "A smear never quite washes clean from a public character," wrote Wallace Stegner. "Cope had no interest in Powell's general plan, probably knew little about it. But his narrowly paleontological and personal attacks could damage everything that Powell had been working for. How much, Powell could not tell until he faced the committees of both houses in the spring."

Although the establishment had stood stoutly by Powell and Marsh during the *Herald* articles—it would have looked foolish if it hadn't—the journalistic depth charges had stressed delicate connections in their steering apparatus. The U.S. Geological Survey was a fragile craft under the best of circumstances, and the damage soon became apparent in the zone of greatest stress, Washington.

Two years earlier, Powell had prevailed on the administration to withdraw subhumid and arid lands west of the 102nd parallel from further settlement until he had surveyed and mapped them as part of his general plan for cooperative use of irrigated land. This farsighted measure would have prevented a great deal of waste and misery, but it went against almost every Western impulse and institution, from the railroads and towns hungry for a steady diet of immigrants, to the speculators and large landowners poised to snap up abandoned homesteads. As an attempt to regulate private property for the public good, Powell's reforms also transgressed many impulses of the United States at large and stirred up powerful Eastern enemies as well. In 1890, a drought intensified the pressure for short-term solutions to the problems of aridity in place of Powell's long-term answers, and Western unrest gushed into a Congress that knew about the *Herald* articles even if the members hadn't read them.

"They were laying for him when he appeared before the House Appropriations Committee in the beginning of June," Stegner wrote. "Somewhere in the room, too, was an echo of Cope's charges of power-grabbing and incompetence, an air of personal mistrust that Powell had rarely met before." Despite Powell's impassioned arguments, Western senators amended an appropriations bill to reopen the withdrawn lands to settlement, and to cut most of Powell's appropriations for his hydrographic surveying and mapping. This setback didn't affect Powell's leadership of the Geological Survey, but it destroyed any immediate possibility of the scientifically directed settlement he'd dreamed of. Stegner called it "the major defeat of his life, and the beginning of the end of his public career." The defeat encouraged Powell's enemies, led by Senator (Big Bill)

Stewart of Nevada and Representative Hilary Herbert of Alabama, to go to work on the survey itself.

One of the survey's strengths was that its budget wasn't itemized; that had prevented congressional enemies from attacking individual appropriations. Now, Stewart and his allies insisted that Powell itemize the budget in 1891. "That was more than ominous," Stegner observed; "it amounted to a vote of no confidence. It meant that whispers about the scientific Tammany had found hearers." Hilary Herbert had already picked an item to attack. Powell had printed a forty-four-page abstract of Marsh's 1880 *Odontornithes* monograph, and Herbert regarded this account of an evolutionary link between reptiles and birds as a bizarre waste of taxpayers' dollars. *"Birds with teeth!"* became the slogan of Powell's congressional enemies, who circulated a deluxe edition of the monograph as evidence of the survey's extravagance.

Marsh's Eastern eminence had obscured Western dislike of his debunking reformism, but it had deep roots, as shown by the Kansas publication of William Webb's "Professor Paleozoic" caricature twenty years earlier. "Again, as in the newspaper scandal of January 1890," Stegner wrote, "it was Marsh whose flank had turned to expose Powell's position . . . Herbert then expanded that opening into a full public airing of all the whispers and slanders of 1890, 1885, 1878–79, 1874, and all the years between."

It didn't matter that printing monograph abstracts was standard government practice, that Marsh had paid for the deluxe edition himself, or that *Odontornithes* was one of the century's great scientific advances. *"Birds with teeth!"* worked where Cope's and Endlich's 23,000-word docket had failed in 1885. Congress amended the 1892 appropriations bill to cut the Geological Survey's budget by about a fourth, and to eliminate most of the funding for paleontology. The *Tribune* called the defeat "unexpected," but the *Herald* scoffed at "the usual attack and defense of the survey and Major Powell," reporting with relish that Congress had cut over $55 million from the budget, and that "the biggest cut was in the Sundry Civil Bill," Powell's bill. "That brought the house down," wrote

Stegner, "and with it much of the structure of government science that Powell had labored with for more than twenty years."

On July 20, Powell telegraphed to Marsh: "Appropriation cut off. Please send your resignation at once." It must have been a humiliating shock. According to Schuchert, Marsh had been "kept fully aware of the dangers that threatened," but was "somewhat at a loss to understand the attacks on his work." He had been using his own congressional contacts to try to control the damage; he'd even explained the *Odontornithes* monograph's financing to Hilary Herbert personally. But none of it worked this time. To be fired so peremptorily because of a great scientific *achievement* must have seemed a monstrous injustice. Powell tried to take some of the sting out of it two weeks later with a letter granting Marsh the title of Honorary Survey Paleontologist, but he evidently didn't try very hard.

"I do not think that the Director intended that the letter of dismissal should be sent to you as it was," Powell's assistant Charles D. Walcott wrote to Marsh that fall, "as I cannot conceive, after all the support that he has given your work, that he would intentionally place you in this position." Walcott promised to ask Powell why he'd been so brusque, but there is no evidence he did so or that Marsh and Powell associated afterward. Powell would not have been human if he hadn't regretted choosing so quarrelsome an ally back in 1878. He himself left the survey in 1893 and spent the years, until his death from a stroke in 1902, writing theoretical geology and philosophy as though sick of politics, scientific or otherwise.

"Marsh's world may well have seemed to be crashing down about his ears that July morning in 1892," wrote Schuchert. "Not only was it a severe blow to his pride to be summarily dismissed from the commanding position he had held for a decade, but the quite unexpected removal of $4000 from his annual income coincided with a 'low' in his other finances. One of his biographers said of Marsh that he planned his scientific work as if he planned to live forever. It might equally be said of him that he spent his money as though he expected the source of it to be 'eternal.'" The Gilded Age was

Punch*'s 1890 cartoon of "Ringmaster Marsh"*

ending, however, and the recession that would sink Clarence King in 1893 dried up most of the Peabody estate funds that had supported Marsh's princely life. The Professor had to apply to Yale for a salary and to mortgage his house to the university to pay for its upkeep. What was even more damaging to his self-esteem was that he had to curtail his fossil-collecting, "and from that time," wrote Schuchert, "he acquired materials only through an occasional purchase or from short field trips."

Having been called a plagiarist, charlatan, and tyrant in the newspapers inflicted a subtler impoverishment. There would be no more articles starring the Yale Professor as fearless reformer, eagle-eyed investigator, or incorruptible witness. Celebrity declined toward notoriety, as with the paunchy "Ringmaster Marsh" of the *Punch* cartoon on September 13, 1890. A friendly caricature, it still portrayed its subject as the bedizened leader of an animal act, a considerable decline from the well-dressed colossus who fifteen years earlier had scorned the U.S. secretary of the interior in *Frank*

Leslie's Illustrated Weekly. Other caricatures were less friendly, as when the *Herald,* for reasons obscure even by Bennett standards, used a fossil feud sidebar to embroider its full-page hype (headlined WAR! WAR! WAR! WAR!) of a prizefight between "Gentleman Jim" Corbett and "Lanky Bob" Fitzsimmons in Carson City, Nevada. BLOODY CARSON, read the sidebar head beside an engraving of two vaguely uintathere-like creatures at each other's throats: DREAD SCENES WERE THOSE.

"In the beautiful valley of Carson," the *Herald* informed its boxing fans, "within recent geologic time, but ages ago as man's life is reckoned, there roamed huge aggressive monsters whose very tread shook the earth, and whose hoarse and frightful voices would have raised the hair, in mortal terror, of such innocent little bruisers as the puny pugilists who will battle for supremacy next Wednesday. Think of Mr. Big Tailed Little Headed Megatherium Americanum, sixty feet in length and weighing a few odd tons, out on the warpath for Professor Heavy Voiced Curve Horned Mylodon, of equally terrific proportions when stripped and the tiny baby prize fighter of the modern day sinks into utter, absolute insignificance." The article added that the California Academy of Sciences had investigated twenty-two-inch-long footprints of a "giant primitive man, or troglodyte," in the area, as well as gigantic saurian bones, but that "Professor Marsh" had "ruined what might have been a good story" by identifying the footprints as those of a *Mylodon,* a giant sloth. "Who cares about Carson Valley, buried in the misty haze and sediment of geologic time?" the *Herald* concluded.

Even Marsh's scientific luster dimmed to a degree. Age was decimating the evolutionist establishment that had exalted him: Asa Gray died in 1888, Huxley in 1895. The new generation was skeptical of the Darwinism Othniel had championed, and it included more enemies, like Baur and Scott, than friends. Alfred. S. Romer wrote that "the old-timers I knew when young (Cope and Marsh had died long before) all liked Cope and almost to a man they hated Marsh's guts." Ernest Howe, a Yale geology graduate of the 1890s, began an unpublished Marsh biography in the 1930s. He remembered his subject "not as a scientist or partisan, but as a rather pompous but kindly old gentleman who had hunted buffalo in the dim past."

Marsh evidently tried to ignore the decline. His friend, the English paleontologist Arthur Smith Woodward, said the *Punch* drawing "delighted" him. If that déclassé attention was welcome, he must have recalled the days of Red Cloud and the Huxley lectures with sad nostalgia. Smith Woodward once accompanied Marsh to a luncheon at which none of the upper-class guests "appeared to have heard either of Professor Marsh or of his discoveries." He recalled, "I shall always remember the disappointed expression of Marsh as we left on our return to London."

In two years, fun-loving fate had reduced Marsh from scholar-princedom to a condition closer to Cope's than either man might have thought possible in 1889. By the summer of 1892, each was a university professor with an international scientific reputation, but without much wealth, fame, or political power, and with a certain notoriety. To put a crowning touch on this artful symmetry, fate awarded Cope a post as paleontologist on the Texas Geological Survey in 1892, allowing him to resume major collecting just as Marsh's money troubles forced him to stop.

Jack-in-the-box Cope popped up again almost as though the precipitous 1880s had never been. "I feel cheerful," Edward wrote to his wife from a westbound train in May. "In packing up I made a great haul in the lower trunk of the closet under the stairs. I found saddle bags, canteen, fish net, gum blanket, no. 2 shoes, and a good coat which I took along." Soon he was on the West Texas Staked Plains, a region he'd "often wanted to see . . . as it furnishes the connecting link between the geology of the center and the Gulf-Atlantic regions." The Texas prairie exhilarated him the way Kansas had in 1871. "I never saw such a variety of beautiful flowers," he wrote from Espella, Dickens County. "I never saw so many prairie dogs, as it is one continuous town from Big Springs here (100 miles) . . . The weather is like that of the northern plains, only *more so*. It is blazing hot by day and cold at night."

Cope looked for Permian reptiles but with "very poor success," so he moved on to Triassic badlands and found a few things, including part of "a carnivorous dinosaur distinctly related to *Laelaps*." When he started digging in Tertiary badlands around a hill called Mount Blanco, however, he discovered another unknown slice of reality, a segment of the late Pliocene epoch just before the Ice Age. Paleontologists still refer to this time as the Blancan. "I find the formation and the species it contains fill a gap in the series of formations and connect an earlier with a later fauna," he wrote to Annie on June 5. The site yielded giant tortoises, mastodons, camels, ground sloths, and horses. The camels and horses were all "new to science," including *Equus simplicidens*, the species I would dig for at Hagerman Fossil Beds in 1997.

"I have seen so many new and strange things that I seem to have been absent a long time," Edward said, "and am affected considerably with homesickness." In July, though, he moved on to the South Dakota Sioux Reservation, and in a sense brought the saga of Western frontier paleontology full circle. That was where it all had begun half a century before, even though Cope himself had not been there.

"The trip involved some risks," he told Annie. "The Sioux have been angered lately by trespassers on their reservation who have stolen their horses and cattle, and they are very suspicious of white people who want to go to their land. They have a way of killing stray people whom they don't know . . . A man who went to gather fossils would be a mighty subject of speculation, with a good deal of doubt as to that being his real object." Mosquitoes proved more troublesome than Indians. Less than two years after Wounded Knee, the exhausted Plains tribes had given up fighting, and Cope would suffer nothing worse than bumpy wagon rides as he collected in the Dakotas during the summer of 1893 as well as in 1892. The Lakotas treated him hospitably on several occasions, although he gave most of the credit for this kindness to missionaries, "good New England types," like a Congregationalist named Miss Collins. "So between hands the Indians have every inducement to be good," he wrote to Annie smugly in July 1893. "An Indian department which feeds them and schools them; missions to convert and school them; and an Army to whip them if they are bad. It is no wonder they are generally good at present."

Not that the Dakotas were entirely subdued. On July 23, 1892, Cope described the remains of a large Catholic church, dedicated only two weeks before, that lay "strewn over the prairie" after one of the frequent violent storms. "Many were the narrow escapes from flying timbers and lightning that were told me," he wrote to Julia. "The quarters of the employees and the guard house, each say 75-foot-long buildings, had been lifted from their foundations and carried 20–40 feet. A store was wrecked by flying timbers that pierced its walls. If I had been caught in camp in the hurricane, I would have had a bad time." When he went to Cretaceous formations along the Grand River in northwest South Dakota to look for dinosaurs, thunderstorms brooded almost continually over his diggings. Lightning "played across the sky in forked streams or occasionally descended to the ground in blinding bolts," he wrote. "Storms raged on three sides of us, and appeared certain to reach us from one or the other quarter . . .

"As it grew late," he continued, "we turned down a low hill to the left and climbed a low bench at the foot of an opposite hill. I saw a low bare bank and lying about white objects . . . The Sioux boy motioned me to him and showed me a low clay hill ⅛ mile S. We went to it, passing over fragments of bones all the way. But at the hill were numerous bones of giants nearly entire; one could hardly walk without stepping on them. Presently I stopped before a curious object buried deeply in the ground and beheld the nearly entire skull of a great reptile related to *Hadrosaurus,* some 3½ feet long . . . The mound in the center of the bone bed is just the shape of a grave, say 500 ft. long by 50 wide and 30 high; a fitting tomb for the family buried beneath and around it. It is an eerie-looking place with a poisonous alkaline pool just below."

Cope tried to dispel the eeriness with condescension. "The lightning did not avenge the disturbance of the bones," he added, "and we dug them up and boxed and shipped them . . . I remembered the Indian legend of the place as narrated to me by Miss Collins. The Sioux knew of it long ago, but they believed that the bones belonged to evil monsters which were slain by lightning by the Great Spirit. They would not touch the bones for fear a like fate would befall them. So they were fortunately preserved for the more intelligent white man who is not troubled by such superstitions." His sense of superiority seemed to soften, however. A year later he retold the same myth rather as William Bartram might have narrated it. "The Indians believe that the fossil bones are those of huge serpents which burrow in the earth," he wrote to Annie, "and that lightning is always trying to find and kill them, and that those bones we see have been so killed. The bad lands have been made by the efforts of lightning to find them in the earth." The story may have come to seem less a childish superstition, and more a metaphor for the precariousness of life on the plains—indeed, of life in Philadelphia.

Cope never went west again. The burst of energy that followed his recovery from "Marsh fever" spent itself after 1893, and he had plenty to occupy him at home. He'd become a popular teacher and lecturer, and along with his relentlessly prolific scientific writing, he "spread himself" on such social issues as gender and race, espousing conventional notions of white male evolutionary superiority. Such writing paid fairly well, unlike scientific papers, and Cope needed the money. In May 1894, he wrote to Persifor Frazer, "It is very possible that within three months I may not have any place to work in, nor any place to store my collections."

His financial woes finally eased when the American Museum of Natural History bought his fossil collection after Osborn became curator of paleontology. Its first purchase, in 1895, was of North American mammals, and it got a bargain, according to Cope, who had wanted $50,000 for them. Osborn recalled an embarrassing encounter: "As he left President Jesup's office with a rather downcast look [Cope said], 'Osborn, I have sold my collection for $18,000 less than it is worth. I have ceased to wonder why Mr. Jesup has been so successful in business!' " (Morris K. Jesup, the museum's millionaire director, had tried and failed to acquire Marsh's collections in the 1880s.) Two years later the museum bought the rest—North American reptiles, amphibians, and fishes; South American mammals; and European vertebrates—bringing the total price to $60,550, substantially less than the $75,000 Cope said he'd spent to amass them.

Cope had wanted his fossils to go to the Philadelphia Academy, but it couldn't or wouldn't buy them. "No disappointment was greater to him," wrote Frazer, "than that of being obliged to see the best part of his collection of a life-time pass away from his native city and the society in whose halls he had performed most of his work." Still, Edward must have been relieved to unload his tons of bones after a decade of trying. The sale further evened the score with Marsh, since his life's work now also reposed in a great institution instead of cluttering indifferent Philadelphia's attics and warehouses.

In 1895, the American Association for the Advancement of Science elected Cope president, settling yet another score with perennial National Academy of Sciences President Marsh, who stepped down from his office the same year. "Although I spoke to nobody I was elected on the 1st ballot as though all had been 'preordained,' " Edward happily wrote to Osborn. "Perhaps it was!" In 1895, four years after Joseph Leidy's death, the University of Pennsylvania promoted him to Leidy's former position as chair of zoology and comparative anatomy, the university's closest equivalent to Marsh's Yale paleontology chair.

Private life also preoccupied Cope, although less agreeably. In the spring of 1894, his daughter Julia left home to marry William H. Collins, a professor of Astronomy at Haverford College. Edward and his wife moved to Haverford to be near her, but this arrangement apparently didn't work. Around that time, according to Osborn's 1930 letter to the Princeton University Press editor, Annie "left him . . . largely for financial reasons," but also "from finding it too hard to live with a genius." Osborn added that he didn't think she had left from "finding one too many snakes in her shoe," perhaps intended as a joke about Cope's herpetology but interpreted by Jane Pierce Davidson, in light of the letter's sexual allegations, as another reference to his philandering. Whatever the reason, Cope mainly lived alone in the office of his 2102 Pine Street lab for the rest of this life, another symmetry with Marsh's no longer splendid loneliness in New Haven.

Marsh also seems to have ended his Western travels by the mid-1890s, although his secretiveness makes it hard to tell. That trait deepened with age and, as Schuchert wrote, "led him to extraordinary lengths to protect his fossil treasures from what he regarded as prying eyes." The table of contents in his unpublished memoirs cited five Western trips between 1880 and 1890, but Marsh never got around to describing them. There may have been others after 1890, particularly if Marsh really crossed the Rockies twenty-seven times, as he said. Except for some collecting of Mesozoic plants, though, the robber-baron part of his career ended in 1892, when he had to tell John Bell Hatcher to stop acquiring ceratopsian bones and start cataloguing the ones he had. Marsh would not have been amused later in the decade when Osborn sent the young ace collector Barnum Brown to Como Bluff. After nosing around for a summer, Brown happened on another bonanza of Jurassic sauropods at a nearby ridge called the Medicine Bow Anticline, where bones were so thick that a sheepherder had built a cabin with them. Maybe Othniel reassured himself that no one would get much that he hadn't, and that he'd already described most of it anyway.

The Professor certainly had more than enough fossils to occupy him. Schuchert appraised them at over a million dollars, even after $200,000 worth owned by the U.S.G.S. had gone to the Smithsonian in three railroad cars after Marsh's death. Marsh acquired his *sobriquet* "the Great Dismal Swamp" largely through his unwearying efforts to persuade fellow clubmen to contribute to a new museum wing for his overflowing bone pile. In the 1930s, Peabody employees still complained about the mass of unprepared and unopened material in the storerooms. S. Dillon Ripley, then the head curator of the Peabody, told *The New Yorker* in 1962 that he was *still* opening Marsh boxes. "Recently, for example, we found the jaw of an early Tertiary bat in one of his cases," Ripley said. "Very rare indeed."

Observing at an alumni dinner that he couldn't take his fossils with him, and that they'd burn up where some people thought he was going, Marsh donated his collection to Yale in 1898. "In extinct Mammals, Birds, and Reptiles, of North America, this series stands pre-eminent," he boasted in the Deed of Gift. "This collection was pronounced by Huxley, who examined it with care in 1876, to be surpassed by no other in the world. Darwin, in 1878, expressed a strong desire to visit America for the sole purpose of seeing this collection. Since then it has more than doubled in size and value, and still holds first rank." So much for the American Museum of Natural History, by gad.

Marsh spent most of the mid- to late 1890s classifying and restoring his earlier finds, beginning in 1891 with the unfortunate *Brontosaurus* and proceeding to more lasting success with *Triceratops, Stegosaurus,* and other famous genera. He made several European trips to coordinate Old and New World dinosaur classification, and did restorations of four European genera, including *Iguanadon* and *Megalosaurus,* the first giant saurians to have been called dinosaurs by Sir Richard Owen in 1841. His bipedal restorations of these two were a great improvement over the quadrupedal ones that Owen and artist Waterhouse Hawkins had made in 1853, and Marsh did not hesitate to say so at an 1895 meeting of the British Association for the Advancement of Science.

"The dinosaurs seem," he said, "to have suffered much from both their enemies and their friends. Many of them were destroyed and dismembered long ago by their natural enemies, but, more recently, their friends have done them further injustice by putting together their scattered remains and restoring them to supposed lifelike forms . . . We now know from good evidence that both *Megalosaurus* and *Iguanodon* were bipedal, and to represent them as creeping, except in their extreme youth, would be almost as incongruous as to do this by the genus *Homo.*"

Courtly Old World scientists didn't hold such Yankee brashness against the Yale Professor. In the fall of 1897, the French Academy of Sciences gave him the Cuvier Medal, the highest paleontological award, for "the most remarkable work either on the Animal Kingdom or on Geology." Marsh was the third American to get the prize, following Louis Agassiz and Joseph Leidy. "This recognition must have been doubly sweet," Schuchert wrote, "coming as it did in the midst of so many difficulties." Fate, however, denied Othniel one sweetness of this crowning honor. Edward Cope was no longer alive to envy it.

19

Death

In the mid-1890s, Cope "laughingly" told Henry Fairfield Osborn that his doctor had warned him to "avoid horseback riding and exposure to water," but that "his health had been greatly improved in the course of a summer by three hundred miles' exercise in the saddle in North Dakota and several weeks' wading in the North Jersey swamps." That was typical of his attitude to the care of his health, but it obviously had its limits. In the spring of 1896, he developed an illness that he couldn't laugh off or suppress with self-administered drugs; it was a flare-up of urogenital infections that had plagued him at least since his 1872 Fort Bridger collapse, when he'd written to Ferdinand Hayden that his ailments included "slight attacks of orchitis, cystitis etc.," as well as fever.

The cause of the infections isn't clear. Cope's womanizing reputation and a lack of medical records gave rise to talk that he had syphilis and had accidentally killed himself by trying to treat it with formalin, a chemical used to preserve specimens. There's little evidence for the rumor, however. The "carbuncles" Cope had complained of at Fort Bridger could have been symptoms of syphilis but could have had many other causes. Cope, in later life, showed none of the nervous impairment associated with advanced syphilis. A lesser venereal disease, such as gonorrhea, may have contributed to his urogenital troubles, but, again, there's no record of one, and Cope made no secret of the troubles, mentioning them in letters to his wife and daughter. In any case venereal disease is not required to account for his last physical breakdown. What is impressive is that he remained ambulatory for as long as he did, considering his hectic life, his chronic malaria, and the powerful toxins he used to treat it.

In an April 28 letter, Edward wrote to his wife that he was too ill for a planned visit to Haverford, and that his university secretary, Anna Brown, was bringing him food. "The Dr. says that if I will keep quiet I will escape a serious sickness, so I am *keeping still,*" he continued impatiently. "Probably by tomorrow evg. I will be able to come out." A week later he wrote to a scientific associate that he was convalescing and reading a Joseph Leidy book on Florida fossils, his impatience apparently having overcome the doctor's advice. Cope couldn't afford to rest, since his income depended on his teaching, lecturing, and writing. Osborn wrote that "even severe pain did not stop his work nor keep him from his university classes."

He resumed his normal pursuits during that summer and fall. After a visit to the American Museum in September, he complained to Osborn that few of the fossils from his collection were labeled as his, although the bill of sale had specified it. "I am not particular about it on the score of vanity," he added, doubtless to Osborn's amusement, "but some reasons why it should be done have occurred to me, so that I am inclined to insist on the carrying out of the letter of the paper." Edward may have sensed that he had little time, since he began negotiating to sell the rest of his collections, and started winding up his affairs generally. "I am back in Washington and am now very near the end of my work here," he wrote to his daughter on August 9, 1896. "I have finished very nearly the final review of the geographical distribution of the reptilian squamata [snakes] of North America." As Osborn observed, "Cope was old and weary at fifty-six; he had borne the heat of battle and wanted rest."

In February 1897, Edward again wrote to Annie that he was confined to his bedroom-office in the care of his secretary. (Annie also was in poor health that year, suffering from "stomach or chest trouble," according to his letter; she would live until the 1930s.) The renewed attack of "cystitis" laster longer. "For two weeks I have been confined mostly to the house on Pine Street, and for three days I was in bed with the same trouble that sent me there last spring," he wrote to his sister on March 10. "I have had a great deal of pain, but am much better of it now, and certain dangerous symptoms which I had have disappeared. I have to be very careful however lest they return, so am still a prisoner . . . Miss Brown, my assistant, is an excellent nurse, and Dr. Slocum will pull me through as far as care can do it. One of these days, however, I will probably have to go to a hospital and stay there for a time for an operation."

Cope's "dangerous symptoms" probably included the inability to urinate, because of his inflamed prostate gland and urethra, and that may have been what motivated his desperate turn to self-medication. On March 16, Osborn wrote that he was "aghast at hearing of your use of formalin and do not approve of it at all. I trust you will find some less hardening medium of remedy and relief." Osborn arranged with "the most experienced surgeon in New York" for the operation Cope needed, and even reserved a private hospital room for him, but "Cope for the first time in his life showed a lack of physical courage" and hung back. Surgery for advanced prostate disease at the time generally entailed castration, so his reluctance was understandable.

Edward simply was unable to give up his active, self-reliant life. Whenever he felt a little better, he went on some excursion and promptly had a relapse. "I did not suffer any material damage from my exercise at Overbrook, nevertheless, I do not seem to get well, and remain rather stationary," he wrote to Julia on March 15. Ten days later, he told Annie: "Third day noon I was taken with one of those nervous paroxysms and it lasted me until 8 or 9 A.M. 4th day morning. I took an enormous quantity of morphia and finally resorted to belladonna. At the time mentioned the blockade was broken and the agony ceased . . . I took no more morphia until last night when I was seized with a most vigorous chill so I took a little morphia which broke it up . . . This being my experience of the last three days, I do not know much about anything else." On March 27, he wrote: "I had a good night last night, and am feeling pretty well excepting that I am weak . . . The sky looks beautiful out the window, and I dare say that in a few days the country will be charming. I am anxious to get out, but cannot yet awhile."

Cope was not so ill that he couldn't connive in a last swipe at Marsh. One of the era's best magazines was the lavishly illustrated *American Century,* which was liberal in outlook, highly literate, and not given to scandal-mongering. Somehow, though, probably through Osborn's growing influence, the dying jack-in-the-box and his shady journalistic henchman wormed their way into those lofty premises. When the November issue was published, seven months after Edward's death, it contained a nine-page article by William Hosea Ballou, "Strange Creatures of the Past: Gigantic Saurians of the Reptilian Age," much of which sounded more like Cope than Ballou. It provided a sound if long-winded view of contemporary knowledge about dinosaurs, from their origin in the Triassic Period to their Jurassic climax and Cretaceous extinction. Osborn contributed a six-page introductory piece about Edward's career, wherein he called his late friend not only "a great naturalist" but "certainly the greatest America has produced."

The features included elaborate illustrations by Charles Knight, the young artist who had described Cope as having "the most animal" mind he'd encountered. Knight had been working on dinosaur reconstructions at the American Museum of Natural History, and his illustrations were almost photographic, showing Triassic *Dimetrodon,* Jurassic *Stegosaurus,* and Cretaceous *Hadrosaurus* in lifelike poses against vivid landscapes. They must have had an enormous impact on readers who had never seen such things so realistically rendered before. The double feature probably was the nineteenth century's best popular presentation of dinosaurs and the beginning of the twentieth's fascination with them.

The feature's effect on *one* reader would have been enormous in a different way. Ballou's article was calculated to infuriate Marsh, since it gave the clear impression that Cope had discovered, named, and classified North American dinosaurs single-handedly. "In his laboratory in Philadelphia," it began, "the late Professor Edward Drinker Cope devoted many years to the study of the fossil or petrified skeletons of the gigantic saurians, or lizard like reptiles . . . During several months preceding his death his original and interesting views upon these animals and his ingenious speculations regarding their habitats, were imparted to the writer." From there, it was "Cope discovered this" and "Cope theorized that" throughout the piece. Aside from a few parenthetic species names, the only mentions of the Yale Professor were an acknowledgment (probably by a *Century Magazine* "layman editor") that Knight's painting of a ceratopsian captioned *Agathaumas* was based on "Professor O. C. Marsh's prior reconstruction of *Triceratops prorsus*," and that Marsh had been the first to discover *Stegosaurus*. Even those mentions would have infuriated Othniel, since they implied that Cope had done as much as Marsh to describe the two dinosaurs. Ballou also persisted in using Cope's beloved but invalid name *Laelaps*, although Marsh had changed it to *Dryptosaurus* twenty years earlier.

Cope may have used *Laelaps*, indeed, as a parting gesture toward Marsh. The twenty-three-year-old Knight drew heavily on the dying paleontologist's advice in preparing the illustrations. (Osborn had introduced the men in a New York hotel room.) Edward seems to have had the lifelong feud in mind when he talked to the illustrator about the "devourer of *Hadrosaurus* and all else it could lay claws on," because Knight's rendering is odd. Carnivorous dinosaurs usually are shown attacking prey, but the two in his vivid painting are fighting. One, jaws agape, leaps high at the other, which lies on its back with its claws extended. The scene recalls a titanic cockfight, and the grinning saurians give an impression of rabid joy. A volcano erupts in the background. Cope may have meant the prehistoric squabble as a last "happy warrior" joke to accompany the Ballou article's posthumous nose-thumbing, although stolid Othniel surely wasn't amused.

Knight's 1897 painting, based on Cope's ideas

Cope even may have thought he might encounter Marsh beyond the grave. On March 27, he sent an optimistic letter to his Aunt Jane, who had helped raise him after his mother's death. "I don't expect to leave this world yet for awhile," he wrote, "but I shall do so when the time comes with a full belief that it will be a change greatly for the better." This avowed intimation of personal immortality may have been merely another attempt to placate his Quaker relatives, yet it could have been genuine. Cope certainly never adopted Huxleyan agnosticism, and his religious ideas remained highly developed despite his move away from Christianity. "He became very reticent about speaking about God," wrote Osborn, "but his conviction that God made the world, and that right and wrong existed very tangibly in it, remained unshaken."

Edward did not have long to wait for the "change." A March 29 letter to Osborn asking that he be allowed to postpone surgery until he recovered from his "last relapse" was the end of his voluminous writing. He remained lucid for a while, and Osborn found him bedridden but "bright and animated" during a visit on April 5.

"After a casual reference to his suffering," Osborn wrote, "he at once entered into a discussion of my views on the origin of the mammalia." But Cope's long-suffering kidneys had failed, and uremic poisoning had set in. "He was confined to his cot-bed, on all sides of which fossil bones were piled," Osborn wrote. "It was considered unwise to move him and on April 12, 1897, he died there.

"Not many days before his death," Osborn continued, "Cope is said, according to Persifor Frazer, to have delivered in his delirium a lecture on the Felidae with all his charm of manner and diction and all his profound knowledge of the history of the subject, the discoveries up to date, and their relations to each other and to other great problems of zoology." Jane Pierce Davidson threw a grimmer light on the last days: "I have not been able to determine who, if anyone, was present at the actual time of Cope's death on April 12."

Edward's funeral was well attended. Indeed, he had two—one at Pine Street for his scientific friends, one at his father's estate for his family and Quaker friends. Osborn went to both, and left a heartfelt description of the first. Climbing "for the last time the steps of 2102 Pine Street, [I] found the coffin placed on Cope's two study tables and covered with a dark cloth, upon which lay a spray of white magnolia blossoms and green leaves. Chairs on each side of the coffin seated five of Cope's personal scientific friends and colleagues . . . I took another seat facing the coffin and we all sat in perfect Quaker silence for what seemed an interminable length of time. There was no sound except for the slow migrations of a land tortoise from Florida, which had been one of the living pets of Cope's study from some time past. This tortoise seemed to feel that something was wrong and rendered an humble herpetological tribute by wandering about as quietly as possible.

"But the most real reptilian memory and tribute was displayed in a glass vivarium not far from Cope's study table, in which a Gila monster, *Heloderma suspectum,* slowly circled the glass walls of his cage. It may have been my imagination, but every time the circle brought him nearer the study table, he seemed to rise on his fore-limbs and gaze steadily at the magnolia blossoms on top of the coffin. I then recalled that I had often seen my friend Cope rise and scratch the top of the head of his Gila monster as it circled about the enclosure; perhaps the *Heloderma* was looking once more for this friendly hand." Finding the Quaker silence unbearable, Osborn stood up and read a passage from the Book of Job, the one that begins: "Where wast thou when I laid the foundations of the earth? Declare, if thou has understanding. Who hath laid the measures thereof, if thou knowest? Or who hath stretched the line upon it?" Then he said, "These are the problems to which our friend devoted his life."

The American Naturalist ran a lavish memorial issue in May with a two-page, black-bordered death notice by Persifor Frazer, a six-page obituary by the morphologist J. S. Kingsley, and six portraits (a bust, a painting, and four photos). Frazer wrote that Cope's main contribution had not been "species-making" but recognizing the significance of his many discoveries to the whole structure of science. Kingsley called him "a peer of Huxley and Owen" and said he had opened up a new field of paleontology with his first trip to the West in 1870. Neither mentioned Marsh or the *Herald* controversy, but both alluded to them. "He had no patience with the view that it is honest, that it is honorable, to hire others to do intellectual work," commented Kingsley. Frazer wrote that Cope's "views and convictions on all subjects were impersonal and raised far above the malarial atmosphere of jealousy and malice." He predicted that the effect of Cope's passing on the world of science would not become evident until European journals reported it, because "there is not so general an appreciation here as abroad of the services he has rendered to Natural History."

In fact, American reporting was light. The National Academy's *American Journal of Science* ran a one-page obituary that ludicrously erred on Cope's vital statistics, saying that he died at forty-six, and that "his published works large and small are said to exceed 350 in number." The popular press did somewhat better. The *New York Times* obituary on May 13 used the paltry 350 publications figure, but it balanced error with error by granting Cope "a full course of Medicine" at the University of Pennsylvania. Otherwise, the quarter-column piece was fairly accurate, describing Cope's "fossil investigations" as "of great value to science" and crediting him with the discovery of 155 new vertebrate species during his Wyoming and Colorado expeditions in 1872 and 1873. "The dead scientist was a constant investigator, an intelligent reader, and a worker of extraordinary capacity," it observed, and then perceptively summed up Cope's role as leader of the neo-Lamarckians: "In the preface of his first collected philosophical essays, *The Origin of the Fittest,* Professor Cope asserts that the important point is not only the survival but the origin of fitness, and this he traces to the inheritance of individual reaction to environment."

Not surprisingly, the *Times* obituary failed to mention Marsh or the *Herald* controversy. Indeed, the *Herald* ignored them in its own slightly longer obituary. Its piece, one-third of a column, included a portrait and a thorough enumeration of Cope's offices and honors, but it was less perceptive than was the *Times* in discussing his work. "Professor Cope took an active part in geological explorations in New Jersey, Maryland, North Carolina, Pennsylvania, and the Western States, resulting in the discovery of more than 600 new species of vertebrates," it curtly declared. "His scientific writings were voluminous and mainly technical in character."

There's no record of Marsh's reaction to Cope's death, though the news must not have been unwelcome. Jack-in-the-box would bounce up no more. Marsh would not survive Cope for long, however, which is suggestive of the lengthy marriages wherein one mate soon follows the other into the grave. If Marsh hated Cope, he also may have felt more strongly about him than he did about anyone else during most of his adult life. When such ideas seemed important, a Freudian analyst might have made a case for the youthful Marsh sublimating a latent homosexual attraction to the vibrant Philadelphian into a lifelong project of professional domination. Even without Freudian analysis, it's easy to surmise that Cope's disappearance would have left a gap in his enemy's emotional life. Habits of feeling are hard to break, and hating the dead can be as dispiriting as loving them.

It's not as though Marsh died of a broken heart. Cope's passing evidently had no immediate effect on his fine living and his flair as a *raconteur.* "I have had enough of it all and feel much as a small boy does the day after the 4th of July," he jauntily wrote to his neighbor, Professor Brush, from the 1897 International Geological Congress in Moscow. "You will soon get the particulars of it all, and I will not risk spoiling the story by trying to condense it here. Some of it will not soon be forgotten." A photograph taken at a European garden party during the trip looks less than jaunty, however, and Marsh arranged to leave his collection to Yale that winter. He was in poor heath during a European trip the next summer. Schuchert wrote that he had "been ill during that year with arterial trouble so that he was no longer able to walk readily as he had done all his life." Like Cope, Marsh died from the way he had lived. Arteriosclerosis lurked in the Gilded Age's gala banquets.

Marsh's growing debilitation perhaps explains why he never got back at Osborn and Ballou for their dinosaur articles. He certainly tried. "His threat of some time ago to get even with you and me for the *Century* article seems to be moving slowly," Ballou wrote to Osborn on February 11, 1898. "When the *Century* wrote me some time ago that it had made an arrangement with Marsh for articles on extinct animals, I promptly filed a sharp protest against deceiving the public with such articles and called attention to the fact that Marsh was under certain grave charges affecting the accuracy of his work and its authorship. Today I had a talk with the *Century* people and was informed that the *Century* had so far failed to accept any of Marsh's work."

Ballou added that Marsh also had called on the editor of *Popular Science Monthly* to demand that he not print a Ballou article on marine reptiles, "as I had grossly misrepresented his work and had done so in the *Century* article." The editor had replied "that the *Monthly* had a habit of printing what it pleased and preferred my work to his and would print it," and had added that "my *Century* article was very fair to Marsh and that Marsh's only complaint could be that so much space was given to Cope." Marsh "went away in a huff."

Ballou was in a megalomaniacal mood. In January, he'd written to Osborn that Marsh's work would "be simultaneously attacked everywhere, by arrangement," and another February letter proclaimed that he had "opened fire on Marsh" and started a congressional investigation "as to specimens belonging to the United States" at Yale. "I am wholly familiar with the Washington crowd and their methods, and the men who interfere with this process this time will be decapitated," he bragged. "This will be no easy-going exposé such as I made of Powell and Marsh in the *Herald* some years ago, where both sides were allowed to have their say." Ballou even credited himself for Marsh's releasing part of his collection to the Smithsonian and donating the rest to Yale. "I think he will be compelled to turn over all to the government," he wrote on February 11, "which paid him and his employees for their work."

Osborn's response to these gradiosities was suave but pleased. "I have been much interested in your various letters with the items of news concerning the Scientific Progress to which you have given your leisure time," he wrote. "Professor Marsh seems to spend a great deal of his time complaining of his being misrepresented, but he appears to enjoy very little sympathy, a fact I think is becoming pretty widely understood."

In fact, Marsh had a great deal to complain of. On December 20, 1897, the *New Haven Register* reprinted a *New York Sun* article about the impending death of George Hull, the Onondaga Giant fabricator, without mentioning Marsh's starring role in the affair. Evidently in response to complaints, the December 26 *Register* featured an original piece: "EXPOSED BY PROFESSOR MARSH: Yale Man Tells of Cardiff Giant Episode: History of a Gigantic Fraud." It reprinted Marsh's famous letter and gave the impression that Othniel had exposed Hull's creation single-handedly, since it didn't mention the hoax-debunking by Professor Boynton, which the *Herald* earlier had reported. "Professor Marsh read the Story of the Finding of the Figure," a subhead proclaimed. "And After Inspection Quickly settled the Claims of its 'Finders' of its great Antiquity."

This solo revival of the professorial Sherlock Holmes evidently caught a glowing eye. A day later, the *New York Herald* carried a short piece headed REMAINS OF FOUR BEINGS ARE FOUND EMBEDDED IN STONE. Datelined "Sandstone, Minnesota," it reported that "human beings turned to stone and now forming a part of the rock have been found in the big sandstone quarries here, as well as copper utensils of a bygone age, showing that this section was once inhabited by a people antedating the Indians." The *New York Times, World,* and *Tribune* ignored the exciting discovery, but their silence didn't discourage the *Herald* from running a much longer, illustrated piece on Sunday, January 2. PETRIFIED REMAINS OF A RACE OF GIANTS appeared above a large drawing of workmen loading two blocky humanoid forms onto a railroad car.

"Scientists who have had an opportunity of inspecting the finds say it may have been that the bodies were buried and in time the rocky formation grew up around them," the *Herald* reported with a turgid nonsensicality worthy of its 1890 feud features, "but if this is a fact, how was it that the bodies did not decay in the natural course of time, for the rock formation is evidently the result of scores of years of slow growth?" Bennett's zombies went into considerably more detail on the "giants" that Sunday, but then dropped them, leaving a distinct odor of the Central Park Zoo. It seemed very much like another kick at the Professor's crumbling pedestal.

Marsh had threatened to kick back in his memoirs, written around 1890. "I of course regretted that the Antelope Pre-Adamite man turned out to be a myth and the high antiquity of our ancestors lost so promising a support," he had written of the Nebraska "human remains" he'd debunked in 1868. "How many similar discoveries are recorded, and fade away each decade of the closing century, may be gathered by the instances mentioned in the later pages of this work. The far West, so prolific of strange things, has furnished her full share, but the East is not far behind, and for pure invention and public credulity takes the lead." He never got around to finishing his memoirs or to specifying the "further instances," and however much the *Herald*'s PETRIFIED REMAINS must have vexed him (the 1897–1898 articles are among his papers at Yale), he evidently lacked the energy to debunk them.

Another round of journalistic monkeyshines a year later came as a climax to the ironies that had pursued Othniel since 1890. Probably prompted by the *Century*'s lavish articles, "dinomania" finally erupted in the sensationalist press. On December 11, 1898, Hearst's *New York Journal* published a full-page Sunday spread with a banner headline proclaiming, MOST COLOSSAL ANIMAL EVER ON EARTH JUST FOUND OUT WEST, and a photo of a handlebar-mustached individual, identified as "Bill Reeder," standing beside a six-foot thigh bone. In fact, "Reeder" was the Como Bluff collector William Harlow Reed, who, tired of sheep, had become the curator of the University of Wyoming Geological Museum and had discovered a new sauropod in Utah. The article spurred Hearst's yellow press rival, the *World*, to run its own dinosaur piece two weeks later. This one, headlined NEW YORK'S NEWEST, OLDEST, BIGGEST CITIZEN, MR. C. DINOSAUR, was about a Wyoming *Camarasaurus* that Henry Fairfield Osborn's diggers had found, probably at Como Bluff, although Osborn, characteristically, didn't say.

Marsh may have taken some pleasure in the *Journal*'s sensationalizing of his brain children. It featured an illustration of his *Brontosaurus* reconstruction as well as of his faithful collector Reed. His enjoyment, however, would have been brief. The *Journal* matched the *Herald*'s BLOODY CARSON boxing sidebar in clamorous stupidity. WHEN IT ATE IT FILLED A STOMACH LARGE ENOUGH TO HOLD THREE ELEPHANTS, screeched the Hearst organ. ITS TERRIBLE ROAR COULD BE HEARD FOR TEN MILES. A picture of a biepedal sauropod peering pensively into an eleventh-floor window of the New York Life Building illustrated the paper's claim: WHEN IT STOOD UP ITS HEIGHT WAS EQUAL TO ELEVEN STORIES OF A SKYSCRAPER.

The *World*'s piece would have pleased Marsh even less, since it completely ignored his work as it labored to top the *Journal* with multiple bipedal sauropods feeding on palm trees and a *Camarasaurus* skeleton towering beside a trolley car. HE LIVED FOR FIVE HUNDRED YEARS, squawked the Pulitzer calliope. HIS APPETITE WAS LITERALLY INSATIABLE; HE ATE EVERY WAKING MINUTE! In the face of such embarrassments, it perhaps was as well that paleontology's Sherlock Holmes had three months to live.

Marsh's last days are virtually undocumented by contemporary letters or memoirs. His secretiveness explains this, and it would not seem remarkable except in contrast to the documentation concerning Cope's. One effect of this silence, justified or not, is to suggest a life ending in emotional inanition—exhausted, isolated blankness. Schuchert's 1940 account seems to support this. The only companions it mentions are Peabody Museum attendants.

"The last of February, 1899," Schuchert wrote, "he made a trip to Washington—not a pleasant one, doubtless—as Walcott was pressing him to close up his Survey connections and send material on to Washington . . . Returning to New Haven next morning, he walked from the railway station to the New Haven House through heavy rain, and Bostwick and Westbrook [museum aides], summoned thither, found him wet through and chilly. His illness really started from that time, although he insisted on coming to work at the Museum each day, driving around to the back door in a cab and having the men take him up to his office in the freight elevator.

"One day," Schuchert continued, "Gibb [Marsh's fossil preparator] found him as usual when he went up at noon to make his tea, but at three-thirty Bostwick came rushing down to the basement with the word 'the Professor's sick.' Gibb went upstairs immediately and Marsh stretched out his hand toward him, saying 'Gibb, I'm sick,' but at the same time put his finger to his lips so that Gibb should tell no one else. Gibb took him down in the elevator and helped him into a cab. Asked if someone should not go home with him, he shook his head, lay back in the seat, and dropped his hands beside him. Gibb got into the cab and begged to be allowed to go along. Marsh again said, 'No,' then shaking Gibb's hand, said, 'Good-bye, Gibb,' and was driven off. He never left his home again and in less than a week (March 18) was dead of pneumonia. After services at the Battell Chapel, with a memorable eulogy by Professor George P. Fisher of the Yale Divinity School, he was buried in the Yale plot in Grove Street Cemetery."

Schuchert concluded that Marsh's "death was the occasion for widespread comment by the press and few were the newspapers that did not carry some account of his colorful career or some one of the countless 'tall stories' that had grown up around his name." Indeed, the scientific establishment did Othniel proud in the *American Journal of Science,* with a photo and Charles E. Beecher's nineteen-page memorial essay placing him among "leading men of science in America." Beecher wrote that Marsh "brought forth in such rapid succession so many astonishing things that the unexpected became the rule," and concluded, "One cannot help being impressed by their signal brilliancy, their great number, and especially by their unique importance in the field of organic evolution. Were all other evidence lost or wanting, the law of evolution would still have a firm foundation in incontrovertible fact."

"One smiles a little to think how all this fanfare would have delighted the subject of it!" reflected Schuchert. Yet in the popular press, Marsh's faded public persona seemed to blur his scientific fame. He would have been less than delighted at his *New York Times* obituary of March 19. It started out promisingly with a stack of headlines—lacking in the *Times*'s obituary of Cope—and an effusive introduction, calling him "among the greatest investigators and scholars of the age . . . his reputation on account of his marvelous achievements in paleontology being worldwide." When it came to describing the achievements, however, the obituary became surprisingly garbled. It credited Marsh with discovering "two new orders of large mammals from the eocenetertiary of the Rocky Mountains, the Tillodontia, which seems to be related to the carnivores, ungulates, and rodents of elephantine bulk bearing on their heads two or more horn cores." The obituary made better sense of the Red Cloud affair, but then relapsed into confusion with an apocryphal True West anecdote that had Marsh, while "digging Indian mounds in the Black Hills," encounter an apparently "wild" Sioux who turned out to be a Yale Divinity graduate. It was as though a distracted editor had concocted the piece from a Yale press release and clubmen's hearsay.

The *Times* and the *American Journal of Science* again ignored the 1890 *Herald* feud, but this time the *Herald* didn't ignore it. Its March 19 obituary for Marsh was a bit longer than the one for Cope, and carried headlines—Cope's had none—proclaiming his "Services to Science" and "Many Perils in Exploration of the Wilds of the West." But it described those offhandedly: "His trip to the Rocky Mountains disclosed the richness of the fossil treasures there, and in 1870 he succeeded in organizing the first of the Yale scientific expeditions, which he led himself." It disposed of the Red Cloud affair just as curtly; Marsh's investigation had led to "a cessation of many abuses."

The *Herald* then lurched into the 1890 scandal rather as a bored tippler will blurt out some anecdote of questionable taste. "He had a controversy also with Professor E. D. Cope of the University of Pennsylvania in which his scientific attainments were disputed. The argument was ended in 1890, when on January 19, in the columns of the *Herald,* Professor Marsh published an entire page devoted to the refutation of his critics." Since the obituary didn't say that the critics *had* been refuted, this was a deft turn of journalistic equivocation, leaving the reader to wonder about the authenticity of Marsh's "scientific attainments." It was as though the Commodore had to throw one last monkey wrench into the works of scientific reputation before he went on his merry way.

Bennett's way, in fact, was merry in 1899 despite a lifetime spent in abusing society in general and his liver in particular. Only a year younger than Cope, he was no advertisement for the perils of sociopathy and alcoholism—not for the rich, anyway. His paper's circulation passed half a million that year because its Spanish-American War reporting was far superior to Hearst's and to Pulitzer's. In 1900, he used his record profits to buy a steam yacht, which he named the *Lysistrata* "for a Greek lady reputed to be very beautiful and very fast," and cruised about the Mediterranean with the crowned heads of Europe. ("Say Bennett!" a drunken staff artist once shouted. "How much do you pay these kings for riding around with you?") In 1902, when an envious Hearst wired him: "Is the *Herald* for sale?" the Commodore cabled back: "Price for *Herald* three cents daily. Five cents Sunday."

But Bennett was tempting fate with this egregious mockery of a serious rival. Hearst didn't forget, and after the *Herald* helped defeat his run for governor of New York in 1906, he did an exposé of its sleazy "personals," which got the Commodore indicted in federal court for pandering. Bennett had to pay large fines and stop running the "personals," and the *Herald* never recovered from the loss of advertising revenue and the dive of its circulation, which plummeted from 511,000 to 60,000 in a decade. Shorn of his wealth, Bennett underwent a personality change, becoming a "Scotch miser," as one of his correspondents put it. He stopped drinking, became an Episcopalian, and in 1915 married Maude de Reuter, the widow of the news agency's founder. From then on he was "a thoroughly domesticated man, dotingly attached to the lovely Maude."

Even so fortunate a being could not prevail forever. Bennett's plans for a giant owl mausoleum in New York had collapsed after another crazy millionaire, named Harry Thaw, murdered his architect, Stanford White, for having a love affair with Thaw's wife, the chorus girl Evelyn Nesbit. Considering this a bad omen, Bennett contented himself with having owls carved on an otherwise unmarked gravestone in a small cemetery in Passy, near Paris. To the end, the Commodore retained a strange power. He expected to die of a stroke on his seventy-seventh birthday, as his father had, and he more or less did. On that day, May 10, 1918, he suffered a massive brain hemorrhage, to which he succumbed a few days later.

Rivals published respectful obituaries, as though fearing the monster even in death. "There was a deeper sagacity in this man than was generally realized," said the *New York Times*. "He had his own way through a long life . . . but he was not an idler. What editor worked harder? He got out of life what he wanted. A vivid figure, such as a novelist could not invent, and yet that seems too fanciful for actuality. He was a sort of Fairy Prince to the last."

20

The Skeleton Drummer

Most tragedies end in death, but the bone-war story does not. O. C. Marsh has rested quietly in the Yale cemetery this past century, no doubt fulfilling the wishes of the academic and scientific friends who put him there. E. D. Cope's professional friends got a surprise when they attended his second funeral, at Fairfield. "At the close of the long but silent service," wrote Osborn, "the coffin was removed and I was expecting to accompany the body to the grave, but instead Cope's son-in-law, Professor William H. Collins, touched me on the shoulder and said: 'Friend, in the next room thee will find something of interest.'" Osborn followed Collins into an adjoining room, where, behind the closed door, Cope's will was read.

The will, which left Cope's small estate to his wife and daughter, contained a codicil directing that "after my funeral my body shall be presented to the Anthropometric Society and that an autopsy shall be performed on it. My brain shall be preserved in their collection of brains, and my skeleton shall be prepared and preserved in their collection, in a locked case or drawer, and shall not be placed on exhibition, but shall be open to the inspection of students of anthropology." Osborn learned that "in response to an invitation issued to a number of anatomists to leave their remains for the advancement of scientific knowledge, Cope had resolved only a year before his death to join his friends, Doctor Joseph Leidy and Doctor John A. Ryder, in the final tribute of all his earthly possessions to the cause of science." Cope asked that the rest of his body be cremated and his "ashes be preserved in the same place as shall contain the ashes of my esteemed friends, Dr. Jos. Leidy and Dr. John A. Ryder."

The *New York Herald* carried a Sunday feature on the Anthropometric Society a year later. "There are three hundred men in various parts of the United States who have bequeathed their brains to science," it began. "These men are neither cranks nor fools. They represent the cream of American professional life. Science with them is paramount. But nevertheless, it is a grewsome organization, this American Anthropometric Society." The feature reported that Dr. William Pepper, "the eminent surgeon and physician," had founded the society so that anatomists could have the brains of "learned men" to work on. (It didn't identify Pepper as the tormented University of Pennsylvania provost of its 1890 feud coverage.) It noted that Joseph Leidy's brain was the first to be acquired, in 1891, and that his brain was "unusually heavy and richly convoluted . . . the most interesting of all the specimens in the society's possession."

It continued, "Of the members who followed Dr. Leidy there is one who was more interesting in many respects than he. The late Professor Edward Drinker Cope, the foremost paleontologist of the country, whose unequalled collection of fossils the Metropolitan Museum made strenuous efforts to secure, bequeathed not only his brain, but his skeleton, as well, to the scientific world . . . The disjointed bones of the great scholar lie in a rough box in the basement, where they have remained untouched since the day which saw the completion of the process of maceration. Dr. Cope's skeleton is of no more value . . . than that of any other man, so far as concerns scientific study, and it is not likely that it will ever be mounted for exhibition. It is interesting, however, as showing the enthusiasm of the man, who, after giving the best of his life to science, bequeathed to it after death not only a goodly share of his wealth, but also all of his body that could be of any possible value." This echo of the 1890 *Herald*'s "plucky Pennsylvanian" sob story ran beside photos of Cope and of Pepper, who had just died of a "breakdown" in California. The reporter added that both brains showed "a thickening of the arterial system indicative of the extremely active nervous systems of these . . . eminent men."

Cope's final request may seem surprising for a man whose dying letter to his aunt had expressed hope of life after death. His religious ideas didn't include anything as literal as bodily resurrection, however, and there was nothing strange about leaving one's corpse to science. As the *Herald* fulsomely proclaimed, it was a soberly idealistic thing to do at a time when opportunities for anatomical research were limited. Cope was unusual in leaving his skeleton to the Anthropometric Society as well as his brain, but he always did more than others. Brain and bones both figured in at least one scientific study, by Professor Edward A. Spitzka, in 1903. An Anthropometric Society founder, Spitzka made a project of examining eminent scientists' brains, including that of John Wesley Powell, which he pronounced "superior" although showing signs of senile atrophy. He thought Cope's brain manifested more potential for imaginative theorizing than did Joseph Leidy's, but less for rational observation. Of Cope's skull he wrote, "The specimen is remarkable on account of the proportionately large size of the cranium as compared to the face, in this respect approaching the notable skull of Kant."

Later attitudes toward Cope's remains, however, were less scientific. Fun-loving fate plucked his bones from the basement and dispatched them on undignified, sometimes bizarre wanderings, like the skeleton drummer who leads the *danse macabre* in medieval woodcuts. A tangled body of legend grew up around them during these travels.

Perhaps too "grewsome" for the new century, the Anthropometric Society was dissolved in the early 1900s, and its collection passed to its parent institution, the Wistar Institute of Anatomy and Biology. The Wistar Institute kept Edward's brain, but in 1966 gave his skeleton, labeled Specimen 4989, to the University of Pennsylvania's Museum of Anthropology, where it fell into the hands of an evolutionist as complicated, in his way, as Cope was. Loren Eiseley, then the Benjamin Franklin Professor of Anthropology, saw Cope's name on carton 4989 and decided to take it into his office for "safe-keeping."

"Professor Eiseley removed the bones gently and respectfully from the dust-covered box and laid them out on his long conference table to be sure that all of Professor Cope's skeleton was there," wrote his assistant, Caroline Werkley, in a 1975 *Smithsonian* magazine article. "Then, believing that the former paleontologist, who had himself stored many a bone in a carton, would feel more at home in the box, returned him to it, his skull balanced properly on top, as should be." Werkley's article described Cope's career and feud with Marsh, but focused on a sentimental attachment that Eiseley formed toward the skeletal paleontologist. A large photograph showed him hovering over Edward's boxed bones.

"He is such an integral part of the office," Werkley wrote, "that when Eiseley and the director of the museum lunch together in the Benjamin Franklin Professor's office, they have been known to offer a solemn toast to Skeleton No. 4989 as they drink their sherry . . . Cope is decorated festively at Christmastime with tinsel and holly by the office staff. He has a bunch of dried flowers tucked into his carton in the other months of the year. Professor Eiseley once bought him a birthday present, a printing block of a skeleton seated before a desk shrouded with cobwebs and spiders."

It's hard to imagine what Cope would have thought of all this. If he'd read Eiseley, he might have been amused. Few essays evoke the eerie, "cemetary" aspect of the Western fossil fields as vividly as "The Slit," wherein the narrator climbs down one of the watchful badlands holes and finds himself face to face with a Paleocene epoch primate: "The skull lay tilted in such a manner that it stared, sightless, up at me," Eiseley wrote, "as though I, too, were already caught a few feet above him in the strata and, in my turn, were staring upward at that strip of sky which the ages were carrying farther

Eiseley took Cope's remains to his office for "safekeeping"

away from me beneath the tumbling debris of falling mountains. The creature had never lived to see a man, and I, what was it I was never going to see?"

Yet Eiseley's care of Cope's remains was not without mishap. His biographer, Gale E. Christiansen, wrote that he lost the skull "sometime in the mid-1970s," after an artist from the Museum of Natural History borrowed it as a model for a bust. When, in 1997, I asked Alan Mann, curator of the University of Pennsylvania Anthropology Museum, about this, he responded that there was indeed serious doubt about the 4989 skull then in the museum. "The skeletal bones, which I have every reason to believe are those of Cope, are extremely poorly preserved; they were never properly degreased and they are exceedingly unpleasant to handle," he wrote to me. "This is not true of the skull, which was well prepared and as a result feels and *looks* very different in texture from the post-crania. They simply do not appear as if they came from the same individual."

Later, Mann changed his mind about the skull. "A very careful comparison of the drawings published by Dr. Spitzka with the skull demonstrates *without question*," he wrote in 1998, "that the skull presently identified as that of Cope is indeed the same skull that Dr. Spitzka pictured in his article." He decided that what Eiseley had lost in the 1970s was not Cope's cranium, but his jawbone. "After Eiseley's death, when the skeleton entered the collections, there was no lower jaw associated with it," he wrote. "Yet, in the photograph of Eiseley and the Cope skull in the Werkley magazine piece, there is a lower jaw sitting below the skull. Comparing the skull and lower jaw in the photograph with the drawings in Dr. Spitzka's article, it is clear that the skull is the same in both illustrations, but the lower jaw is not . . . I think that for the photograph, Eiseley, aware of the loss of the lower jaw, hunted in the museum's skeletal collections for another jaw similar to that of the original, and posed with that . . . Afterward, he returned this jaw to the museum's collections."

Whatever Eiseley did with Cope's skeleton, he made a tangle of its curatorship. Werkley's *Smithsonian* article said that he "expressed to some of his colleagues the wistful hope that possibly his friend Professor Cope . . . might ultimately be interred with him." According to Christiansen's biography, Eiseley conspired with his nephew, Jim Hahn, to fulfill that hope. In 1977, "Hahn took the bones from Eiseley's office to the funeral home where his uncle was laid out," Christiansen wrote, but "wilted upon realizing that the mortician soon would discover the stowaway and thwart the plot, leaving him no choice but to smuggle Cope's skeleton back into the museum." With characteristic ambiguity, Eiseley had also encouraged university associates to try to get the skeleton buried in the Cope family grave at Haverford Friends Meeting House Cemetery. "After his death," wrote Jane Pierce Davidson, "his assistant, Caroline E. Werkley, and several officials at the university attempted to arrange this burial."

Burial proved unfeasible, and the bones reverted to the museum. But their travels weren't over, and the story got stranger. During the *Jurassic Park* craze in the early 1990s, a magazine photographer, Louie Psihoyos, decided to collect bone-war memorabilia as part of a dinosaur book project, and learned about Cope's skeleton from the curator of the Philadelphia Academy of Natural Sciences, Ted Daeschler. Evidently eager to compete with a glut of dinosaur novelties, Psihoyos asked to photograph the skeleton, writing in his book, *Hunting Dinosaurs,* that Daeschler "shook his head in disbelief, smiled, and made a few crosstown phone calls to secure permission" from the university. Psihoyos then "wheeled over to the Museum [where] the curator of collections . . . had gone to lunch and left the esteemed professor with a security guard at the front desk. Instructions were left for us to sign a permission slip like the one you fill out when you check out a library book." According to Alan Mann, who was abroad at the time, the loan form stipulated that "the skull would remain at the Academy during its time away from the University Museum," but Psihoyos took the bones away and kept them for years.

He used the bones for journalistic stunts that might have startled even James Gordon Bennett. He carried the skull around in a cardboard box and sprang it on paleontologists he interviewed, strangely reprising Cope's jack-in-the-box role *vis à vis* Marsh. Not surprisingly, those being interviewed often gave responses that lacked substantive reflection. The skull's visit to the Peabody Museum at Yale was so barren of results, except the shocked dismay of John Ostrom, the curator, that Psihoyos was reduced to ghost stories. Camera film disappeared or was mysteriously loaded backward, rendering his work useless. When he photographed Cope before an oil portrait of Marsh, the photos showed a "weird unaccountable blue glow" around the skull. When he photographed Ostrom, a light next to the skull "exploded into flames."

Psihoyos acquired some respect for Cope from the paleontologists. "Everyone wanted to meet and be photographed with him—even in his present condition," he wrote. "Quite to our surprise, we took a back seat to Specimen 4989, and in a strange way Cope seemed to be escorting us, not the other way around." He worried about losing the bones and eventually had a velvet-lined mahogany box made for the skull. Yet he showed less concern for the facts.

He recalled in *Hunting Dinosaurs* that Ted Daeschler told him Cope had left his bones to science because he wanted them to be designated as the type specimen of *Homo sapiens*. When naming a species, taxonomists usually choose a particular specimen as the "lectotype," but the early taxonomist Carolus Linnaeus didn't do so when he named *Homo sapiens* in his *Systema Naturae* in 1758; he merely cited the Socratic injunction: Know thyself. "By actually becoming the type specimen for man," Psihoyos wrote, "Cope arguably would become, taxonomically anyway, the ultimate man and have the ultimate last laugh over his nemesis Professor Marsh." According to the story, science had rejected Cope's request on the grounds that "his bones were badly decalcifying—showing, it appeared, the beginning signs of syphilis." Psihoyos cooperated with Robert Bakker to "nominate Professor Cope" as the *Homo sapiens* type specimen, however, on the grounds that Bakker "had done a search of the current literature and hadn't found either a description or a type specimen for modern man." Psihoyos concluded that "Cope got his wish and was entered into the scientific literature as the elected type specimen for *Homo sapiens*."

Apparently Bakker hadn't searched the literature hard enough. Linnaeus *had* described the human species, and in 1959 the taxonomist W. T. Stearn had selected Linnaeus's remains, buried in Uppsala Cathedral, as the type specimen. When I asked Daeschler about the story he'd told Psihoyos, he said he knew of no basis for it. "It's just something that circulates around this community," he said. "It could be an urban legend." I couldn't find any basis for the story. It would have been just the thing for the *New York Herald*, but the paper's 1898 Anthropometric Society feature made no mention of it. Nor did Edward Spitzka's study, Osborn's biography, or any other contemporary source I saw.

"This is something that has always bothered me," wrote Alan

Mann, "and I know of no evidence for it. Indeed, Cope, being the great scientist that he was, would have known that a specimen lacking the essential defining qualities of the dentition simply could not be employed in such a fashion." Teeth are the crucial feature of mammalian classification, and Cope's and Marsh's dentition studies were among their main contributions. Even if Edward had wanted to be the type specimen, he would have known that his nearly toothless corpse was unsuitable. As to syphilis, Spitzka mentioned no signs of it in Cope's brain in 1903, and Professor Morrie Krikun, of the University of Pennsylvania Medical School, in answer to Jane Pierce Davidson's query in 1995, "stated flatly that he didn't believe Cope had ever had syphilis." And he had examined the bones.

Psihoyos and Bakker weren't the only ones to popularize the Cope legend. In the same 1992 PBS dinosaur documentary wherein Bakker claimed Como Bluff had made dinosaurs "famous overnight," Peter Dodson, professor of paleontology at the University of Pennsylvania, said that Cope had "specified" in his will that his brain be measured, "confident" that science would prove it larger than Marsh's. An atmosphere of amused credulity evidently lulled Psihoyos as he did his stunts. When he scheduled a book-promotion talk at the Philadelphia Academy in the fall of 1994, however, he met an angry response. "It's a tasteless, offensive outrage," said President Keith Thomson in a front-page *Philadelphia Inquirer* article. "They had no authority from us or anyone else to traipse all over the country with the remains of Professor Cope like some kind of college prank."

Psihoyos responded in the *Inquirer* article that he "thought they'd be tickled and honored. We were going to return a little luster and attention to a great but almost forgotten Philadelphia scientist." Instead of luster and attention, however, Cope's reputation got a load of mud. In the same article, the *Inquirer* smeared the old troublemaker, dragging his bones through a lurid version of mid-1990s political correctness. ON THE TRAIL OF A WAYWARD SKULL blared the headline over a photo of Psihoyos brandishing the 4989 cranium: AFTER DEATH A SCIENTIST AND HIS RACIST EGO LIVE

ON. With *Herald*-like bombast, the feature called Cope "a fervent racist, sexist, and anti-Semite," and, under a Mephistophelian portrait, "a Hitleresque racist." The author, Mark Bowden, exaggerated his evidently superficial gleanings from the literature on Cope. As his magazine articles show, Cope was a racist, but he also condemned lynching. And the *Inquirer*'s comparison to Hiter was just the beginning of its distortions.

Bowden spiced Psihoyos's type specimen story with the notion that Cope had "aspired to be recognized as nothing less than the ultimate model of humanity," because he'd written in one essay that a Greek nose "coincides not only with aesthetic beauty, but with developmental perfection." He sweetened the story of Edward's rejection because of syphilis with the notion that science also had rejected him for being "a little on the short side." He then frosted his confection with irony by bringing up the doubts about the identity of skull 4989. "And so, a century-old act of unparalleled egoism receives its poetic comeuppance," he concluded. "If not Cope's, whose skull did Psihoyos have? Nobody knows . . . The skull is of undetermined race, ethnicity, and gender. In other words, more than two centuries after Linnaeus first dodged the issue, Psihoyos may have hit upon the perfect type specimen for *Homo sapiens* after all."

Psihoyos had written: "Professor Edward Cope had been reduced to bone, catalogued, boxed, and shelved like one of his numerous specimens. But in the caring hands of all those admiring paleontologists, one of whom made him the ultimate example of *Homo sapiens,* he seemed to be smiling, as if death wasn't so bad after all." The real Edward Cope, of course, would not have smiled at an invalid scientific status that he had not sought. The only thing about the affair that would have gratified him was that his skeleton finally reverted to scientific custodianship. Psihoyos returned it, according to Alan Mann, after "the museum had asked the FBI to look into its whereabouts." The 4989 bones again reside with the rest of the museum's over five thousand specimens of human skeletal material, where, Mann observed, they have "very limited scientific importance." Cope would not have disagreed. He knew what his scientific importance was.

The story of Cope's posthumous "travels" is an unedifying spectacle of behavior toward the remains of an influential if controversial historical figure. Aside from its silliness, there's an element of scapegoating about it, with poor, vain Edward as the object of obscure resentments to which figures like Darwin and Huxley have long been immune. As Freud observed, the discoverers of evolution's unimaginably ancient kaleidoscope have disturbed humanity's "naïve self-love" catastrophically. There's also a mythic fitness to the circumstance that two men's lifelong bone war should end with the contentious "resurrection" of the skeleton of one. Perhaps the legend of Cope's type specimen request—a legend impossible to substantiate but curiously easy to believe—arose from a sense of that fitness as much as from an impulse to devalue him.

Epilogue: Squabblers on a Raft

Cope's restless skeleton seems another reminder of Jacobean tragedy, and the episode has a perverse relevance to a question I posed at the beginning—whether Cope and Marsh reached some transcendence from their unrestrained rivalry. Being designated the type specimen for humanity might seem a kind of apotheosis, except that Cope evidently neither knew nor cared about it. So what, if anything, did the antagonists learn from their trouble? Cope "knew more" after his slide into poverty, and apparently achieved some liberation from Marsh's eventual fall, though there is no evidence in his writings that he reflected much about this. He may have looked back regretfully on his father's Jeffersonian hopes as he sat alone with his bone hoard, but he didn't say so. It would be exaggerating, at the very least, to suppose that he somehow transcended rivalry by pursuing it so assiduously that it poisoned much of his life.

Marsh suffered less in the feud than did Cope, so we might assume that he learned less. There are, however, hints that he was more reflective than he cared to admit. His alumni speech remark about not being able to take his fossils with him suggests this, as does a sentence at the beginning of *Fossil Hunting in the Rocky Mountains:* "It has been well said that what one truly and earnestly desires in youth, he will have to the full when no longer young, and my own experience thus far has proved no exception to the rule." A sense of glut evidently was not unknown to the aging Othniel. His clumsy attempts at generosity to Baur and the other Germans may have reflected displaced remorse or at least guilt. Still, the flinty Yale Professor expressed no reflections on having suppressed his younger rival.

Each man did attain a legendary status, a tragicomic grandeur, from the feud. If they hadn't quarreled, their scientific achievements would be duly noted and largely forgotten, along with Joseph Leidy's. Perhaps their story will endure as a myth of an age so greedy that men fought over petrified bones, although I wonder whether they would have traded such immortality for happier lives. According to Freud, happiness lies in love and work, and while neither man seems to have had much capacity for happiness in love, both had an extraordinary capacity for work. That capacity did outlast the *Herald* explosion, so perhaps rivalry can lead to transcendence if it leads back from the passions of vanity to the joys of work. But I don't know if that was enough for Cope and Marsh, each of whom did his best work before being consumed by the rivalry. Cope might not have sunk into neo-Lamarckian speculation if Marsh hadn't turned the Darwinian establishment against him. Marsh might not have sunk into torpor if Cope hadn't turned the younger generation against him.

Even less clear than the question of what Cope and Marsh learned from their feud is the matter of its rights and wrongs. Did Marsh crouch in his office copying Oscar Harger's notes into his monographs? Did Cope slink out of museums with bones under his coat? The volume of commentary that these tableaus have generated implies some actuality; the scarcity of evidence for them is equally striking. Nothing that I have read assures me that they are more than elaborate hearsay, like the 1994 *Inquirer*'s confection. Of course, that doesn't make them less compelling. Alfred Romer recalled that Cope was said to have stolen a whale skeleton from Harvard's Museum of Comparative Zoology. The mind reels.

It is often said that history belongs to those who write, but it also belongs to those who gossip. Once a story is planted in oral tradition, it acquires a resilience despite, or even because of, a lack of evidence. Once hearsay has entered the written record—legend mimicking history—it is nearly impossible to dislodge it. I doubt anybody will ever be sure whether Marsh was a plagiarist or Cope a thief. Hearsay not only may have mimicked history; it may have replaced it in the sense that the spectacle of the rivals locked in their eternal combat of plagiarism and theft is more compelling than the shadowy facts. Even the legend of Cope's syphilis-foiled bid to become the taxonomic "ultimate man" may endure simply because its mad-scientist theme is so mythic.

Whatever its actualities, the bone war still concerns us because, near the end of what has been called "Gilded Age II," we are its virtual contemporaries. As Thomas Beer wrote in the 1920s of the 1890s: "Joseph Leidy . . . the foremost of American natural historians . . . was not news value, but any rogue who announced his faith in his mother's Bible and his unaltered trust in the plain people whom he fleeced or cajoled for votes was news value, and is to-day." If Cope had a "poetic comeuppance," it wasn't from Psihoyos's pranks or Bowden's distortions, but from the fact that *Hunting Dinosaurs* and ON THE TRACK OF A WAYWARD SKULL are precisely the journalism of stunt and smear that he conjured up against Marsh. Evolutionary science remains a sideshow of "news value," and some journalists are as ready as Commodore Bennett was to make a travesty of bonehunters' escapades. At one end, Mark Bowden's *Philadelphia Inquirer* article disdained "the growing cult of bone digging." At the other end, creationism's efflorescence has so broadened the scope of journalistic stupidity that newscasters solemnly "moderate debates" on evolution's factuality, as though Marsh and Cope had not proved it a century ago. Bennett would have laughed.

Of course, a *Herald* can't exploit a Cope and Marsh without their help. Late twentieth-century science also has had its continuities with Gilded Age I, and I've heard about more than one modern feud. ON THE TRAIL OF A WAYWARD SKULL carried a trace of the best-known recent public disagreement between evolutionists. The article probably got its facts about Cope's racism from Stephen Jay Gould's popular 1981 book, *The Mismeasure of Man,* a study of scientific bigotry. Although respecting Cope as a paleontologist and theoretician, Gould deplored his social ideas and cited his 1887 encomium to the Greek nose as an example of racist evolutionism. Gould's book also touched on modern science, and cautioned against a potential for bigoted interpretations of sociobiology, the theory of his colleague Edward O. Wilson.

Gould and Wilson have been sparring publicly for decades. In his autobiography, *Naturalist,* Wilson wrote that Gould had "maintained a drumfire of criticism" of sociobiology from its first publication in 1978. Indeed, one Gould essay, "Cardboard Darwinism," called human sociobiology (which says that genes rather than culture determine many social relationships) "the most peculiar of self-proclaimed revolutions in science," based on "a deeply flawed and now discredited mathematical model." A Wilson book, *The Diversity of Life,* returned the compliment by calling Gould's theory, punctuated equilibrium (which says evolutionary change occurs episodically rather than continually), "a renaming, so to speak, as opposed to a reinvention of the wheel" of species theory, for which "the fossil evidence . . . proved weak."

This disagreement can't be equated with the knockdown-drag-out bone war. The sparring between Wilson and Gould has been mild, compared with that bare-knuckle slugging, and it is unclear from their writings whether they feel any personal animosity. Neither has published *ad hominem* attacks, and neither seems to have threatened the other's position. They've spent most of their careers at the same institution, Harvard University. When radical students poured ice water on Wilson during a seminar, Gould condemned the assault as "infantile." Still, the disagreements have similarities, and these, as well as the differences, say things about evolutionary science in the United States today.

The differences show how science has changed since the nineteenth century. Gould and Wilson, each from the middle class, rose to the high establishment because of their scientific achievements, which would have been harder for them to do a century ago. Each is a "neo-Darwinian," basing his thought on a synthesis of Darwin's natural selection and Mendel's genetics (to oversimplify modern evolutionism), and each has egalitarian attitudes toward race, gender, and other social issues. Such commonalities are fundamental to the two men's careers. Someone like Cope, an anti-Darwinian racist without a doctorate or official position, could not be a full-fledged evolutionary scientist today, no matter how rich and brilliant he was. He would have trouble collecting specimens because of legal and institutional restrictions. He would have trouble publishing in any significant measure.

Another difference between past and present is that Wilson and Gould don't fight about bones; they quarrel about theories. To be sure, this is unsurprising, since neither is a vertebrate paleontologist. (Gould is an invertebrate paleontologist; Wilson an entomologist.) Yet it may be symptomatic of a change from the anarchic accessibility that Daniel Boorstin, in the first volume of *The Americans*, saw as unique to New World science. Even if Wilson and Gould were vertebrate paleontologists, it's doubtful that they would quarrel over actual fossils rather than over theories about fossils. The great, untouched "cemetary" of nineteenth-century America is gone. New fossil fields are hard to find and even harder to get at through the labyrinth of modern institutions and ownerships. This suggests that American science has reverted to Boorstin's European model, with scientists an elite professional class of theorizers. Boorstin perhaps would have agreed. He made no mention of naturalists in the last volume of *The Americans*, which is on the special characteristics of the modern United States.

Yet similarities between the past rivalry and the present one run deep in American history. Wilson came to science through experiences like those which influenced Cope. "I became determined at an early age to be a scientist so that I might stay close to the natural world," he wrote in *Naturalist*. A sense of romantic splendor in

Wilson's descriptions of boyhood Florida adventures recalls Cope's Western letters as well as Bartram's *Travels.* "When I focused on the ponds and swamps lying before me," he wrote, "I abandoned all sense of time . . . In my heart I will be an explorer naturalist until I die. I do not think that conception overly romantic or unrealistic." Like Cope, Wilson approaches fieldwork with an eye for ecosystems and yearns after the unknown. His neo-Darwinian theories resemble Cope's neo-Lamarckian ones in the fertile, headlong way they emerge and proliferate. "I was the naturalist scientist in agreement on the need for strict logic and experimental testing," he wrote concerning sociobiology's origin, "but expansive in spirit and far less prone to be critical of hypotheses in the early stages of investigation. I wanted to move evolutionary biology into every potentially congenial subject, roughshod if need be, and as quickly as possible."

Wilson's early naturalist vocation also resembled Cope's in being linked to a Protestant sectarian upbringing, although his was Baptist rather than Quaker. Each developed a traditionalist's outlook. "At my core I am a social conservative, a loyalist," Wilson wrote. "I cherish traditional institutions, the more venerable and ritual laden the better."

Yet for all the authors' traditionalism, perhaps because of it, the writings of both Wilson and Cope display a certain lack of historical and cultural sensibility. Cope's attitude to contemporary art and literature largely was contemptuous (he called John Ruskin "a monumental ass, with the longest ears that ever decorated an empty skull"). Wilson manifested a degree of anti-intellectualism in *Naturalist,* expressing admiration for explorers, mountain climbers, ultramarathoners, military heroes, but "very few" scientists. "Science is modern civilization's highest achievement, but it has few heroes," he wrote. "Most is the felicitous result of bright minds at play."

Stephen Jay Gould grew up in a world of libraries and museums, one built by men like Marsh. "The fourth floor of the American Museum of Natural History was the shrine, principal magic place, the sanctum sanctorum of my youth," he wrote in one essay. Since his formative experience of nature came through institutions, he is more likely to draw an insight from a zoo or baseball game than a walk in the woods, and he perceives natural history as a civilizing pursuit more than a pioneering one. "Great scientists are embedded in their cultures, not divorced from them," he wrote in another. He has a reformist outlook, perhaps a synthesis of his Jewish background and northeastern Puritan-based traditions, and has developed a public persona not unlike Marsh's, investigating such hoaxes as the Piltdown Man and serving as an expert witness in challenges to laws requiring the teaching of creationism.

Like Marsh, Gould is sparing and deliberate in his fieldwork, with tree snails in the Bahamas. "When I think of paleontology in 1910, with its wealth of data and void of ideas, I regard it as a privilege to be working today," he wrote in 1980. He seeks not a splendid unknown, but discoveries that will help to solve established scientific problems. His best-known theory, punctuated equilibrium, does not attempt to revolutionize evolutionary science's relationship to other disciplines, as does Wilson's sociobiology. It modestly tries to resolve one of the more intractable problems the fossil record poses for evolutionary theory: that few fossils demonstrate the continuous gradual changes predicted by natural selection. Marsh's toothed birds and horse ancestries resolved a related problem: the fossil record's paucity of transitional forms.

Restrained in exploration and theorizing, Gould is lavish in linking science to historical and cultural erudition. "Most of all," he wrote of his early interests, "I recall the impressions conveyed in books." His avid browsing in libraries is not unlike Marsh's enthusiasm for collecting cultural artifacts, and both might be linked to a Judeo-Puritan respect for scholarship. In his essays, if not in his science proper, library research is Gould's main resource, and his zeal for reforming American attitudes to culture and nature largely is fueled by it.

Gould's erudition has had its blind spots. It hasn't encompassed the early American natural history tradition behind Cope and Wilson, and, ultimately, Marsh and himself. In *The Mismeasure of Man,* he dismissed early American naturalists as "a collection of eclectic amateurs, bowing before the prestige of European theorists," and an essay about his Bahamian fieldwork contains a capsule history of great naturalists that leaps from Europe to modern America as though the thirteen colonies had never existed. This indifference to the early naturalists' vision of an unspoiled continent, perhaps combined with Gould's urban roots, generates a degree of anti-environmentalism. "Nothing arouses this ardent . . . New Yorker to greater anger," he wrote in a 1975 essay, "than the claims of some self-styled 'ecoactivists' that large cities are 'unnatural' harbingers of our impending destruction."

Yet Gould has not been alone in his indifference to early American naturalists. Wilson, a traditionalist, also seems to have taken natural history's traditions for granted. At least, his writing has displayed little interest in its American past. His 1998 book, *Consilience,* dwelt on the Enlightenment's scientific significance but mentioned no early New World naturalists except—and only in passing—Jefferson and Thoreau. A century ago, Cope and Marsh also were uninterested in American naturalists, even those still living, like Joseph Leidy.

Indifference to a common history seems a constant of American evolutionist antagonisms, and perhaps throws some light on fate's playful knocking together of heads. Mutual blindness assures collisions. Wilson professed astonishment when Gould and his allies attacked sociobiology as a potential justification for social oppression, and realized that his historical indifference had caught him off guard. "I was unprepared because . . . I am an American rather than a European," he wrote in *Naturalist.* "Had I taken a fatal intellectual misstep by crossing the line into human behavior? I expanded my reading into the social sciences and humanities." Environmentalism seems to have blind-sided Gould in the 1980s, when Wilson became a highly respected advocate of biodiversity conservation. Gould included several essays on that subject in his 1993 collection, *Eight Little Piggies,* and wrote in the preface that "one theme of transcendent (and growing) importance has been absent (and shamefully so) from my writing heretofore." One essay cited Alexander Wilson's and John James Audubon's descriptions of passenger pigeons as "canonical stories of the extinction saga."

Gould and Wilson have shown admirable flexibility in learning from their collisions, but learning from rivals is not always creative; it aggravated the antagonism of Cope and Marsh. Vanity and greed can be such compelling motivations that rivals eagerly undergo the humiliations of imitation in the hope of winning. Since a journalistic and political readiness to exploit the hairline crack in American evolutionary science remains, one can imagine future feuds less benign than the one between Gould and Wilson. The stakes in their rivalry have been lower than those of Cope and Marsh—prestige and popularity rather than economic survival—but, then, they live in good times. The nation's unprecedented twentieth-century prosperity has scattered a lot of crumbs for evolutionists. Yet the "American century" is ending, as the New Mexico paleontologist Spencer Lucas observed when we talked about the possible recurrence of a feud like that of Cope and Marsh. Lucas didn't expect the practice of evolutionary science to become easier. He foresaw fewer professional positions on the raft. As the base shrinks, rivalry can be expected to grow.

So what? Almost everyone who has written about the Cope-Marsh feud has observed that their rivalry stimulated the search for fossils. If the scientific infrastructure gets leaner and meaner again, won't the survivors just work harder? Such similes with technological development may be mistaken when applied to historical sciences like paleontology, however. The squabbles between Cope and Marsh stimulated a lot of activity, but much was of the smash-and-grab variety. Paleontologists might learn a lot about the Jurassic Period if they could see some of Como Bluff as it was in 1870. Like the related fossil-fuel industry, paleontology is built on a nonrenewable resource that should be developed with foresight, not just exhausted as quickly and profitably as possible. To be sure, the majority of fossils still lie unstudied in the ground. But paleontologists can't expect to dig them out with Giant Earth Movers, and if technology finds a cheap, nondestructive way of studying fathom-deep fossils, it would be better applied cooperatively than through some latter-day bone war.

The Cope-Marsh feud didn't just smash fossils; it helped to smash John Wesley Powell's farsighted attempt to develop the West in sustainable fashion. Powell's toppling was accidental, but accidents are inevitable when blind antagonists struggle, and the forces that toppled Powell still live in the West and in Congress, where legislators with creationist constituencies and budget-cutting zeal almost got the U.S. Geological Survey abolished in 1994. What if someone of Cope's abilities and hatreds had been on their side? The 1990s produced at least one Cope-like figure—a brilliant "bone sharp" who was imprisoned as a side effect of a fight over the ownership of a *Tyrannosaurus rex* found on public land. Indeed, squabbles in the 1990s between fossil dealers and professional paleontologists over collecting on public lands reprised both the bone war and the wider Powell-centered controversy about land management. "The so-called wise-use movement, which wants to open millions of acres of public land in the West to logging, grazing, mining, hunting, and motorized off-road sports, has joined the fray on the side of the dealers," observed *U.S. News and World Report* in 1996.

Cope and Marsh probably would not have agreed that each stimulated the other's scientific career. It is likely that they *would* have agreed that the rivalry stimulated their lust to get as many fossils as possible. Envy's sense of annihilation engenders a compensating desire to accumulate tokens of self-importance. This accumulative mania, though, is not the source of scientific discovery. In their case, it led to piles of unopened fossil crates.

Writers on the bone war also have observed that rivalry's vanity, greed, and hatred are basic human traits, and that even great scientists can't be expected to be free of them. There's no denying this, but considering what will be at stake in the next century, one may ask whether science can afford to keep such traits on the raft. "It is an unspeakable misfortune that the character of these two distinguished men was not on a level with their abilities," commented W. B. Scott, "for, had they cooperated with each other in a friendly way, their great achievements would have been even more valuable to science." Scott was one of the wisest of his scientific generation, given to questioning what others took for granted. At least one modern paleontologist has regarded the feud with less than complacency. Cope and Marsh "transformed paleontology to a dynamic science and charged it with a spirit of discovery," Björn Kurtén wrote. "At the same time, the rivalry and enmity between these two eminent scientists is a dark chapter in the history of paleontology."

If science is as close as civilization has come to the truth about life, and if there is some point beyond the tangles of legend and history wherein truth matters, then the evolutionary truth must be much more widely understood. In *Naturalist,* E. O. Wilson stated it eloquently: "When the century began, people could still easily think of themselves as transcendent beings, dark angels confined to Earth awaiting redemption by either soul or intellect. Now most or all of the relevant evidence from science points in the opposite direction, that having been born into the natural world and evolved there step by step over millions of years, we are bound to the rest of life in our ecology, our physiology, and even our spirit. In this sense, the way in which we view the natural world, Nature has changed fundamentally." Paleontology's *memento mori* evidence not only of life's unity but of its mutability and fragility was a fundamental cause of that change.

As perennial polls show, however, most people still believe they are "transcendent beings, dark angels confined to Earth awaiting redemption by either soul or intellect." Dreams of escaping Earth's troublesome "cemetary" remain compelling, even for some who are well aware of the evolutionary past. With their "X-Files" intimations of immortality for a still-squabbling Cope and Marsh, Louie Psihoyos's Peabody Museum ghost stories manifested twentieth-century escape fantasies. Science doesn't promise such things, however, and it seems reckless, at the very least, to use the Earth as though there is someplace else for human beings. If the survival of what we most value in nature, or of civilization, or even of our species depends on dreamers being awakened, it may take more than squabblers on a shrinking raft to do it. Science's tragedy may become humanity's.

Notes

Prologue: Assassination by Newspaper

PAGE

2 "unrelated to the achievements of either man," *American Heritage,* vol. 22, no. 5, August 1971, p. 95.

2 "Even yet, its effects persist," Henry Fairfield Osborn, *Cope: Master Naturalist,* p. 28.

2 "Onward and Upward with Science" piece, Geoffrey T. Hellman. "Go on Investigators! Scrutinize!" pp. 142–76.

2 "ten liveliest debates ever," Hal Hellman, *Great Feuds in Science,* p. 121.

4 "merely passed the time of day," Elizabeth Noble Shor, *The Fossil Feud,* p. 60.

5 "On midnight rides," Richard O'Connor, *The Scandalous Mr. Bennett,* p. 49.

5 planned to be buried, Ibid., p. 259.

6 "His employees simply had to get used to the idea," Ibid., p. 162.

6 "must perforce work to please Bennett," Albert Stevens Crockett, *When James Gordon Bennett Was Caliph of Bagdad,* p. 13.

6 "Thus, for all their contemptuous ill treatment," Don Carlos Seitz, *The James Gordon Bennetts,* p. 237.

6 a rapt description of a murdered prostitute's nude corpse, Richard Kluger, *The Paper,* p. 36.

6 classified "personal" ads, Ibid., p. 183.

6 "The coming of the tiger," *New York Herald,* November 9, 1874, p. 3.

7 "Bennett demonstrated the grasp," O'Connor, op. cit., p. 123.

8 "He could only sponsor them," Ibid., p. 123.
8 in an attempt to confirm rumors, Ibid., p. 119.
8 Not even innocent bystanders, Ibid., p. 152.
10 "It is doubtful that any modern controversy," Wallace Stegner, *Beyond the Hundredth Meridian,* p. 324.

1. Prodigy and Heir

12 "the immediate finger of God," Joseph Kastner, *A Species of Eternity,* p. 59.
12 He was a close friend of Benjamin Franklin, Ibid., p. 56.
12 "The ideal of knowledge," Daniel Boorstin, *The Americans: The Colonial Experience,* p. 168.
13 "No American invention," Ibid., p. 150.
13 "whole men confronting," Josephine Herbst, *New Green World,* p. 1.
13 naked Seminole, William Bartram, *Travels,* p. 107.
13 "I long to be free for pursuits of this kind," Kastner, op. cit., p. 120.
14 In 1812, William and his friends, Ibid., p. 260.
14 "such a museum," Ibid., pp. 153–55.
14 his father, Alfred, belonged to, Jane Pierce Davidson, *The Bone Sharp,* p. 22.
15 "a splendid collection of books within the house," Osborn, *Master Naturalist,* p. 39.
15 "I saw Mammoth," Ibid., p. 40.
15 "His mind reached in every direction," Ibid., p. 39.
15 "We saw some Bonetas," Ibid., p. 43.
16 "I had expected a handsome large room," Ibid., p. 52.
16 he offhandedly mentioned a scientific name for a turtle, Ibid., p. 55.
16 "I traced the stream," Osborn, *Impressions of Great Naturalists,* p. 153.
16 "instituted important modifications," Osborn, *Master Naturalist,* p. 479.
17 "I only wish he would be a scholar," Ibid., p. 82.
17 "She was his refuge," Ibid., p. 48.
17 "I have been hoe-harrowing," Ibid., p. 100.
17 "The whole ground," Ibid., p. 103.
18 "What great university of our day," Ibid., p. 80.
19 He'd read Darwin's *Voyage of the Beagle,* Ibid., p. 84.
19 "Near the entrance," Ibid., p. 127.
20 "The collections from Solenhofen," Ibid., p. 122.

20 "If I know myself," Ibid., p. 136.
20 "That he had an inner consciousness," Ibid., p. 80.
21 "My position here," Ibid., p. 142.
21 "I am very glad to do," Ibid., p. 143.
22 "The adventurer-naturalist," Kastner, op. cit., p. 285.
22 "The enlightened bureaucrat," Ibid., p. 185.

2. Stepchild and Laggard

23 "My journey through New England," Kastner, op. cit., p. 173.
24 Thoreau read Bartram's *Travels,* Henry Thoreau, *Journals,* vol. 2, August 6, 1851, pp. 376–77. Thoreau admired the book but faulted William for describing only the Southeast and not his own home region.
24 "He preferred to roam the fields and woods," Schuchert and LeVene, op. cit., p. 16.
25 records of "events," Ibid., p. 24.
25 His father called him, Ibid., p. 18.
25 the twenty-one-year-old's character, Ibid., p. 20.
26 "I changed my mind," Ibid., p. 22.
26 "foresight and shrewd management," Ibid., p. 23.
26 Othniel's 1854 diary, Ibid., p. 25.
26 and also chanced upon his first vertebrate fossil, Ibid., p. 45.
26 which he named *Eosaurus acadiensis,* Ibid., p. 46.
27 Benjamin Silliman, Sr., Hellman, "Go on Investigators! Scrutinize!" p. 142.
27 THE AMERICAN UNIVERSITY, Schuchert and LeVene, op. cit., p. 28.
27 "He was always very odd," Ibid., p. 37.
27 "strong in body, able to haul a dredge," Ibid., p. 36.
28 "I heartily accede," Ibid., p. 42.
28 Such allegations, Ibid., p. 47.
28 he had "strong hopes," Ibid., p. 75.
28 Silliman enthusiastically responded, p. 77.
29 whether he should specialize in mineralogy, Ibid., p. 49.
29 Silliman advised him, Ibid., p. 51.
29 "The faculty proposes," Ibid., p. 54.
29 "By this appointment," Idem.
29 "an indication of the fashion," Ibid., p. 48.

30 although he complained, Ibid., p. 56.
30 "added greatly to my collections," Ibid., p. 59.
30 "The most inviting field," Ibid., p. 61.
30 "So much for mineralogy," Ibid., p. 59.
31 "If romance really touched him," Ibid., p. 351.
31 ("a very unusual thing," he told Uncle George), Ibid., p. 64.
31 "without salary from existing funds," Ibid., p. 65.

3. Fair Prospects in Dirt

32 "We have spent the prime of our lives," Url Lanham, *The Bone Hunters,* p. 7.
35 "for if one link," Osborn, *Master Naturalist,* p. 12.
36 "At the voice of comparative anatomy," Loren Eiseley, *Darwin's Century,* p. 84.
38 "It has been said that," Kraig Adler, ed., *Contributions to the History of Herpetology,* p. 48.
38 "Professor Cope called on me," *New York Herald,* January 19, 1890.
40 "I thought best to go down," Osborn, *Master Naturalist,* p. 157.
40 "This carnivore, then," E. D. Cope, "The Fossil Reptiles of New Jersey," p. 29.
41 "Flummery there is," Osborn, *Master Naturalist,* p. 144
41 "It was a happy, busy time," Ibid., p. 158.
41 "his first complete synopsis," Ibid., p. 156.
41 The pair found "three new species of Saurians," Ibid., p. 157.
41 "I took him through New Jersey," *New York Herald,* January 13, 1890.
42 "The skeleton itself was arranged," *New York Herald,* January 19, 1890.
43 "my predecessor saw the error," *New York Herald,* January 20, 1890.
44 "I have occupation enough," Osborn, *Master Naturalist,* p. 144.
44 "did not at once turn," Schuchert and LeVene, op. cit., p. 94.
44 Cope, however, was interested in Beadle's fossil, E. D. Cope, *Proc. Am. Phil. Soc.,* 11:419. 1870.
45 "We cannot help wonder," Spencer Lucas, "Cope, Marsh, and the Type Specimen of *Lystrosaurus 'frontosus,'* a Mammal-Like Reptile from the Triassic of South Africa," p. 29.

4. Professor Marsh's Traveling Bone and Pony Show

47 "From the uniform, monotonous prairie," Lanham, op. cit., p. 33.

47 A recent biography called him, Leonard Warren, *Joseph Leidy: The Last Man Who Knew Everything*, p. 5.

48 "You can have no idea," Mike Foster, *Strange Genius: The Life of Ferdinand Vandeveer Hayden*, p. 40.

48 "He soon learned," Osborn, *Master Naturalist*, pp. 23–24.

48 a record of facts," Ibid., p. 160.

48 "For whom are you now collecting?" Lanham, op. cit., p. 35.

50 "It reminded me," Ibid., p. 39.

51 "Their feelings toward us," Ibid., p. 42.

52 "undoubtedly human," H.J.W., "On the Wing in the Far West," *New York Times*, June 22, 1868.

52 "Before we approached," Schuchert and LeVene, op. cit., p. 98. (I attribute quotes from this source to Schuchert in the text henceforth for brevity. LeVene mainly wrote about Marsh's youth and education.)

53 "A hatful of bones," Ibid., pp. 98–99.

53 "the first in a long series," Ibid., p. 99.

54 "apparently belonged to the species," Othniel C. Marsh. "Observations on the Metamorphosis of *Siredon* into *Amblystoma*," p. 374.

54 In an 1868 letter, E. D. Cope, letter of (?) 1868 to Prof. F. Barker (?), Othniel Charles Marsh Papers, Manuscripts and Archives, Yale University Library, reel 3, frames 839–40.

54 Marsh did send photos, Cope, letter of October 5, 1868, Othniel Charles Marsh Papers, reel 3, frames 845–46.

54 "Imminence of Indian Wars," Schuchert and LeVene, op. cit., p. 100.

54 "A good many peddlers," *New York Herald*, August 29, 1868, p. 4.

55 "Altogether it is the most," Schuchert and LeVene, op. cit., p. 343.

55 "the Onondaga Giant," *New York Herald*, November 18, 1869.

55 "There you fastidious fools," O'Connor, op. cit., p. 24.

55 "Jimmy Bennett would serve," Kluger, op. cit., p. 142.

56 "was undoubtedly the most brilliantly edited . . . newspaper," O'Connor, op. cit., p. 101.

57 "which they picked up in fear and trembling," Ibid., p. 100.

57 "special instructions," *New York Herald*, August 9, 1868.

57 "at once widely copied," Schuchert and LeVene, op. cit., p. 343.

58 "some old not much visited Indian mound," *New York Herald*, November 18, 1869, p. 5. An article in the October 25, 1912, *New York*

Sun maintained that the hoax's "eventual exposure" was "erroneously credited by Andrew D. White to Professor Marsh of Yale . . . but the man who really ran the monumental humbug to cover was Dr. John V. Boynton of Syracuse."

58 "Altogether, the work is well calculated," "The Onondaga Giant: Professor Marsh, of Yale College Pronounces it a Humbug," *New York Herald,* December 1, 1869, p. 8.

59 "but, unlike most," Andrew White, "The Cardiff Giant: The True Story of a Remarkable Deception," *Century Magazine,* vol. 64, no. 6, p. 951.

59 "Our immortal naturalist," John Noble Wilford, *The Riddle of the Dinosaur,* p. 22.

59 "trusty pistols and the all-essential knife," Schuchert and LeVene, op. cit., p. 102.

59 "That he found favor in their eyes," Ibid., p. 102.

60 "which became an open sesame," Idem.

60 "They, too, went, saw, and conquered," Harry Ziegler, "The Rocky Mountains," *New York Herald,* December 24, 1870.

60 "tipped me a wink," Schuchert and LeVene, op. cit., p. 103. Charles Betts's *Harper's New Monthly Magazine* article was published in October 1871.

60 "it was probably a fine advertisement," Idem.

60 (It was probably less fine . . .), William F. Cody, letter of December 22, 1874, Othniel Charles Marsh Papers, reel 3, frame 612.

60 "with movements characteristic," Schuchert and LeVene, op. cit., p. 104.

61 "What *did* God Almighty," Idem.

61 "No exploration of this region," Ibid., p. 110.

62 "Endeavoring to control," Ibid., p. 112.

62 "twenty-two daughters," Idem.

62 "about six inches long," Ibid., p. 119.

63 "Yale songs were sung," Ibid., p. 115.

63 "On arriving at the table land," Idem.

64 "tipped me a sly wink," Ibid., p. 118.

64 "Its success was far-reaching," Ibid., p. 136.

64 "The Professor of Paleontology," Ziegler, op. cit.

65 "If Professor Marsh," Idem.

65 "In spite of Buffalo Bill's," Idem.

66 (In August of that year . . .), *New York Herald,* August 2, 1871.

5. The Lone Philadelphian

67 "I confess that during the long winter," Schuchert and LeVene, op. cit., p. 123.

68 "These I measured roughly," Idem.

68 "From a scientific point of view," "The Yale Party," *New York Times,* October 17, 1871, p. 2.

69 "The vast extent of our country," Ferdinand Hayden, introduction, *The Extinct Mammalian Fauna of Dakota and Nebraska,* Philadelphia, 1869.

69 that a younger man, Lanham, op. cit., p. 95.

69 "Not handsome or Christian," Osborn, *Master Naturalist,* p. 169.

70 "Building on the reputation," Foster, op cit., p. 155.

70 "He wanted to discover," Ibid., p. 344.

70 Hayden was the star," Ibid., p. 273.

70 Surely the great West," Ibid., p. 171.

70 espoused theories that increased rainfall, Daniel Boorstin, *The Americans: The National Experience,* p. 232.

70 "The growing financial support," Foster, op. cit., p. 345.

70 The exploration-minded *New York Herald,* "Exploration of the Territories," *New York Herald,* May 10, 1873, p. 8.

71 "I never can rely on him," Osborn, *Master Naturalist,* p. 175.

71 "the train ran close by," Ibid., p. 161.

71 "Marsh has been doing," Ibid., p. 160.

72 "I am very fortunate," Ibid., p. 162.

72 His first fossil hunt, Ibid., p. 163.

72 burned two teamsters at the stake, *New York Herald,* July 29, 1871.

72 "If the explorer searches," Lanham, op. cit., p. 97.

73 "Our Mesozoic periods," Osborn, *Master Naturalist,* p. 166.

73 "animal must have been 75 feet long," Ibid., p. 163.

73 "The flowers we have often heard of," Ibid., p. 161.

74 "Thee mustn't be too sure," Ibid., p. 165.

74 "began to speak of," Charles Sternberg, *Life of a Fossil Hunter,* p. 69.

74 "These strange creatures," Lanham, op. cit., p. 98.

75 "Many other huge reptiles," Osborn, *Master Naturalist,* pp. 165–66.

75 "left them behind," Ibid., p. 163.

6. Babel at Fort Bridger

77 "The history of the life of the successive ages," Osborn, *Master Naturalist*, p. 184.

78 "I would be delighted," James V. Carter, letter of June 2, 1872, Othniel Charles Marsh Papers, reel 3, frame 0165.

78 "entered every field," C. E. Beecher, "Othniel Charles Marsh," *American Journal of Science*, vol. 2, June 1899, p. 405.

78 disputed Cope's right to enter the Bridger field," Osborn, *Master Naturalist*, p. 177.

78 Marsh had tried to get Hayden to promise, Ibid., p. 181.

79 He "started west" on October 8, O. C. Marsh, "On the Dates of Professor Cope's Recent Publications," *American Naturalist*, vol. 7, no. 5, p. 304.

79 This delay seems so incongruous. Walter H. Wheeler's essay, "The Uintatheres and the Cope-Marsh War," has "Marsh, Cope, and Leidy . . . independently collecting" at Fort Bridger "in the summer of 1872." R. W. Howard's *The Dawnseekers*, pp. 218–19, has "the Cope-Leidy and Marsh teams" reaching Fort Bridger "a few days apart" and Marsh sending "a telegram east pointing out errors in Cope's identification" of *Loxolophodon*.

79 I will have every facility," Osborn, *Master Naturalist*, p. 183.

79 "He passed himself off," Ibid., p. 186.

80 "You're right," William Berryman Scott, *Some Memories of a Paleontologist*, p. 231.

80 ingenuously described him as "first rate," Osborn, *Master Naturalist*, p. 186.

80 three kinds of gnats, Idem.

80 "ran the wagon in a ditch, Ibid., p. 187.

81 "in some bodily peril," Cope, *The Vertebrata of the Tertiary Formations of the West: Book One*, Preface.

81 "took his meals at Manleys," Schuchert and LeVene., op. cit., p. 179.

81 Smith had quit Cope's outfit, John Chew, letter of August 16, 1872, Othniel Charles Marsh Papers, reel 3, frame 385.

81 Smith himself wrote to Marsh, Robert Plate, *The Dinosaur Hunters*, p. 133.

81 "I am now so convalescent," Osborn, *Master Naturalist*, p. 190.

82 "No scene ever impressed," Lanham, op. cit., p. 110.

83 "Whe got one tusk," Idem.

83 some of Leidy's discoveries, John Chew, letter of August 26, 1872, Othniel Charles Marsh Papers, reel 3, frame 387.

83 "Marsh and Leidy have obtained it near the same time," Osborn, *Master Naturalist,* p. 191.

83 None of the six names proved valid, Lanham, op. cit., p. 113.

83 "It has been the painful duty," Osborn, *Impressions of Great Naturalists,* p. 144.

84 "a world-famous debacle," Lanham, op. cit., p. 113.

84 "He was as incapable of deceit," Shor, *The Fossil Feud,* p. 276.

84 "except a brief tribute by Marsh," Osborn, *Impressions of Great Naturalists,* p. 143.

85 "Less is known of the experiences of this expedition," Schuchert and LeVene, op. cit., p. 126.

85 "As I glanced at my companions," Ibid., p. 127.

85 "As we rode up the crest," Ibid., p. 128.

86 "My only chance of escape," Ibid., p. 129.

86 "The fortunate discovery of these interesting fossils," Ibid., p. 426.

87 "Nothing so startling," Ibid., p. 427.

87 "a genius for appreciating," Osborn, *Impressions of Great Naturalists,* p. 136.

87 "small specimens," Lanham, op. cit., p. 116.

87 "*O See* Marsh," Cope, addressed envelope, Othniel Charles Marsh Papers, reel 3, frame 884.

87 "to whom I had given," Shor, *The Fossil Feud,* p. 35.

87 "All the specimens," Ibid., p. 36.

88 "Now for all this," Idem.

88 "stuck to me like a leech," Ibid., p. 37.

88 "I have another lesson," Osborn, *Master Naturalist,* p. 179.

88 and predicted that they would "be found," Cope, "The Gigantic Mammals of the Genus *Eobasileus,*" p. 159.

88 "Prof. Cope . . . has made," Marsh, "The Fossil Mammals of the Order Dinocerata," p. 152.

89 and that *Eobasileus* actually belonged to *his* genus, Marsh, "On the Genus *Tinoceras* and Its Allies," p. 217.

89 he included a signed statement from the printers, Cope, "On Some of Professor Marsh's Criticisms," p. 295.

90 "I received nothing of the kind from Professor Cope," Marsh, "On the Dates of Professor Cope's Recent Publications," p. 304.

90 "For this kind of sharp practice," Marsh, "Reply to Professor Cope's Explanations," p. IX.

90 "The recklessness of assertion," Cope, "On Professor Marsh's Criticisms," p. I.

90 "As to the learned professor," Osborn, *Master Naturalist,* p. 182.

7. Marsh the Reformer

92 Nearly every day," Schuchert and LeVene, op. cit., pp. 134–36.

92 "So it seems that," Ibid., p. 137.

93 "afraid that if he told us anything," Henry Farnam, unpublished memoir, dated May 6, 1931, p. 8., Othniel Charles Marsh Papers, reel 26, frames 462–63.

93 "that the whole story," Ibid., frame 467.

93 "when the Professor said," William Edward Webb, *Buffalo Land: An Authentic Account of the Discoveries, Adventures, and Mishaps of a Scientific and Sporting Party in the Wild West,* p. 38.

93 "What folly to suppose," Ibid., p. 111.

93 prophesied that the Great Plains, Ibid., p. 194.

93 "the eminent naturalist," Ibid., Preface. Webb's Cope quotes were from *On the Geology and Vertebrate Paleontology of the Cretaceous Strata of Kansas.*

94 "To develop these two lines of field work," Schuchert and LeVene, op. cit., p. 176.

95 "I have upon my table," Ibid., p. 142.

95 "the origin of the Indian race," Plate, op. cit., p. 103.

95 "could be best managed," *New York Tribune,* April 30, 1875 p. 2.

97 Someone then convinced Marsh. That "someone" was "the officers and Saville," according to Schuchert and LeVene, op. cit., p. 144.

98 "We hadn't gone far," Merrill J. Mattes, *Indians, Infants, and Infantry,* 196–97.

99 One of Marsh's assistants, Lanham, op. cit., p. 90.

100 "I heard a great deal," James H. Cook, *Fifty Years on the Old Frontier,* pp. 196–97.

100 "Gladly would I aid you," James H. Cook, letter of May 1931, Othniel Charles Marsh Papers, reel 26, frame 453.

100 "Then the Professor," Schuchert and LeVene, op. cit., p. 145.
103 "This failure of Red Cloud," Ibid., p. 152.
103 Marsh's discomfiture. According to R. W. Howard's *The Dawnseekers,* p. 227, Marsh went to the *Herald* "early in Spring 1875" with evidence of Indian distress and found an "eager listener" in Bennett, who instructed his editors to give Marsh full cooperation and sent "one of his best interviewers" to Washington. Howard gives no reference for this story, however, and *Herald* articles didn't become pro-Marsh until midsummer.
103 (One issue . . .), *New York Herald,* April 23, 1875.
104 "Possibly his days with the Army," Schuchert and LeVene, op. cit., p. 153.
104 "You alone have the will," Ibid., p. 155.
104 "impetuous but not very practical," Ibid., p. 156.
104 "The Professor with his party," Ibid., p. 155.
105 "He required that all things," Seitz, op. cit., p. 237.
105 "the columns of the *Herald,*" *New York Herald,* August 16, 1875.
106 "more lying in this town," Schuchert and LeVene, op. cit., p. 159.
106 and, in the *Springfield Republican,* Ibid., p. 164.
106 Delano's son and the waiters, Ibid., p. 163.
107 "no important reform of the reservation system," Robert Larson, op. cit., p. 168.
107 (Carpenter had caught bonehunting fever . . .), William L. Carpenter, letter of November 5, 1876, Othniel Charles Marsh Papers, reel 3, frame 120.
107 "Mr. R.C. came to my quarters," Ibid., reel 3, frame 0125.
108 displayed "a grim smile of pleasure," Schuchert and LeVene, op. cit., p. 167.
109 "to break the monopoly on scientific publication," William Goetzmann, *Exploration and Empire,* p. 528.
109 "being used extensively," Lanham, op. cit., p. 217.
109 "a hidebound portion of the intellectual community," Foster, op. cit., p. 5.
109 thrived on controversy, Ibid., p. 351.
110 "He had unlocked the last," Stegner, op. cit., p. 6.

8. Cope the Explorer

112 "Last week I did but little," Osborn, *Master Naturalist,* p. 174.
112 "Hayden only gave me $250," Ibid., p. 196.
113 "I have had passable success," Ibid., p. 195.
114 "Then began again," Cope, "Fossil Reptiles of New Jersey," p. 25.
114 "At this place there is," Osborn, *Master Naturalist,* p. 197.
114 "The result is that the beds," Idem.
115 "*marks* distinctly made," Preface, p. 3, Othniel Charles Marsh Papers, reel 26.
115 "I have to furnish him," Osborn, *Master Naturalist,* p. 170.
115 "Wheeler's and Hayden's surveys," Ibid., p. 171.
116 "It is pretty clear," Ibid., p. 202.
116 "I am only scientific," Ibid., p. 204.
116 "I could not have desired," Ibid., p. 206.
116 "I have just returned," Ibid., p. 207.
117 "I am not sure," Ibid., p. 210.
117 There he saw formations, Ibid., p. 201.
118 "I have now a good outfit," Ibid., p. 211.
118 "The weather is lovely," Ibid., p. 213.
119 the first evidence in North America, Cope, "Synopsis of the Vertebrate Fauna of the Puerco Series," *The American Naturalist,* pp. 208–361.
119 "This Paleocene horizon," Osborn, *Master Naturalist,* p. 198.
119 "the first glimpse of vertebrate life," Björn Kurtén, *The Age of Mammals,* p. 37.
120 no animal fossils, just petrified wood, Osborn, *Master Naturalist,* p. 202.
120 "I have been at work," Ibid., p. 213.
120 Equipped with little more than a mule, Cope, *Tertiary Vertebrata,* p. xxvii.
120 "whose size and powers," Cope, "The Creodonta."
120 "The present camp is," Ibid., p. 214.
121 "I have discovered," Idem.
121 "greatly pleased with my results," Ibid., p. 212.
122 His autobiography, Sternberg, op. cit., p. 18.
122 "Go back with me," Ibid., p. 16.
122 "bound me to Cope for four long years," Ibid., p. 33.
123 "All was excitement," Ibid., p. 63.
123 "There is no risk from the Sioux," Osborn, *Master Naturalist,* p. 221.

123 "Judging from past experience," Sternberg, op. cit., p. 63.

123 "When we first met him," Ibid., p. 75.

123 "I begin to feel like a camper," Osborn, *Master Naturalist*, p. 221.

124 "Since my last," Ibid., pp. 223–24.

124 "Last night," he added, Ibid., p. 225.

124 "One man (Assinibonie Jack) rode several miles to see it," Idem. (Samuel Williston told a similar story about Benjamin Mudge and *his* false teeth in Kansas, Elizabeth Noble Shor, *Fossils and Flies*, p. 59.)

124 "The bed that I came for," Osborn, *Master Naturalist*, p. 231.

125 The fossils' fragility, Sternberg, op. cit., p. 88.

125 from watching doctors set broken bones, Schuchert and LeVene, op. cit., p. 174. The English naturalist Henry De la Beche had started plastering fossils in the 1830s, but the technique had not crossed the Atlantic. David A. E. Spalding, *Dinosaur Hunters*, p. 23.

125 "made it clear that dinosaurs," Schuchert and LeVene, op. cit., p. 375.

125 "We find in the high rocks," Osborn, *Master Naturalist*, pp. 231–32.

125 "I assisted him in digging," Sternberg, op. cit., p. 87.

126 "On the whole," Osborn, *Master Naturalist*, p. 226.

126 "Every animal," Sternberg, op. cit., p. 75.

126 "At night the view," Ibid., p. 70.

127 "Cause, fear of Indians," Osborn, *Master Naturalist*, p. 227.

127 "It seems," Sternberg, op. cit., p. 82.

127 "Yesterday, we moved out," Osborn, *Master Naturalist*, p. 228.

127 "I knew the uselessness," Sternberg, op. cit., p. 90.

128 "We took our strongest horse," Osborn, *Master Naturalist*, p. 229.

128 "So all ended well," Idem.

129 "The uproar it created," Ibid., pp. 230–31.

129 Sitting Bull had crossed, Sternberg, op. cit., p. 98.

130 "I had a standing invitation," Sternberg, op. cit., p. 99.

9. Huxley Anoints Marsh

131 "Get all the bones," Schuchert and LeVene, op. cit., p. 172.

131 "My heart was in my mouth," Ibid., p. 186.

131 "I followed your directions," Idem.

132 "completed the series," Ibid., p. 232.

132 Even the anatomist Richard Owen, Mark J. McCarren, *The Contributions of Othniel Charles Marsh*, p. 41.

133 "I heard a world-renowned professor," Marsh, "Introduction and Succession of Vertebrate Life in America," p. 358. This anecdote is from Marsh's address to the American Association for the Advancement of Science meeting in 1877, in Nashville.

133 "Huxley's Church Scientific," Adrian Desmond, *Huxley: From Devil's Disciple to Evolution's High Priest,* p. 501.

134 "At each inquiry," Leonard Huxley, *The Life and Letters of Thomas Henry Huxley,* pp. 494–95.

134 "My own explorations," Schuchert and LeVene, op. cit., p. 236.

135 "There are few English writers," *New York Herald,* September 18, 1876, p. 4.

135 "having for the next ten days," *New York Herald,* September 23, 1876, p. 5.

136 "loudly applauded," Idem.

136 "Seldom has prophecy," Huxley, op. cit., p. 501.

136 "I had him 'corralled,'" Desmond, *Huxley,* p. 485.

137 "In accordance with your wish," Schuchert and LeVene, op. cit., p. 238.

137 "One evening in London," Huxley, op. cit., p. 6.

138 "Your work on these old birds," Schuchert and LeVene, op. cit., p. 247.

138 "To doubt evolution," Marsh, "Introduction and Succession of Vertebrate Life in America," p. 352.

138 "a bold discoverer," Schuchert and LeVene, op. cit., p. 242.

138 "was listened to with deep interest," Ibid., p. 241.

139 "Both Cope and Marsh," Lanham, op. cit., p. 189.

139 Marsh's reputation began to grow, Osborn, *Master Naturalist,* p. 216.

139 "This law has been epitomized," Cope, "On the Method of the Creation of Organic Types," pamphlet, 1871.

140 "America's first great evolutionary theorist," Stephen Jay Gould, *Ontogeny and Phylogeny,* p. 85. Gould gives a lucid if challenging account of Cope's theories in the context of nineteenth-century biology.

140 "It has quite annoyed me," Davidson, op. cit., p. 172.

141 "It was in violent contrast," Osborn, *Master Naturalist,* pp. 533–34.

141 "it caused Darwin himself," Kurtén, op. cit., p. 28.

141 "On some accounts," Osborn, *Master Naturalist,* p. 243.

141 "one person of importance," Ibid., p. 251.

141 "The only traces," Ibid., p. 247.

142 ("The energy of Cope . . ."), Marsh, "Introduction and Succession of Vertebrate Life in America," p. 378.

142 Cope had earned "the contempt of all the scientific men," Desmond, *Huxley,* p. 715.

142 "Huxley has such influence," Osborn, *Master Naturalist,* p. 247.

10. Dinosaurs and Fate

143 "Gad! *Gad!*" Shor, *The Fossil Feud,* p. 42.

144 "Marsh was never to have," Schuchert and LeVene, op. cit., p. 176.

144 "In fact . . . the boxes," Osborn, *Master Naturalist,* p. 244.

144 "We could extract no information," Scott, op. cit., p. 58.

145 "It was so monstrous," Michael Kohl and John McIntosh, eds., *Discovering Dinosaurs in the Old West,* p. 12.

145 "It is not difficult," Schuchert and LeVene, op. cit., p. 190.

145 "Jones cannot interfere," Idem.

145 the code word for pterodactyl was "drag," Shor, *Fossils and Flies,* p. 82.

145 Mudge rightly called them, Schuchert and LeVene, op. cit., p. 190.

147 "This remarkable creature," Osborn, *Master Naturalist,* p. 256.

147 "There are bones," Schuchert and LeVene, op. cit., p. 193.

147 embedded in mudstone or shale, Ibid., p. 194.

147 "Secure all possible," Idem.

148 "I don't propose," Ibid., p. 195.

148 "We measured one shoulder blade," Ibid., p. 196.

149 "A freshly opened box of cigars," Spalding, op. cit., p. 102.

149 The Hayden and King surveys, Lanham, op. cit., p. 177.

149 He promised to "send a ton a week gotten out good," Schuchert and LeVene, op. cit., p. 197.

149 "Canon City and Morrison," Ostrom and McIntosh, op. cit., p. 9.

149 "precautions to keep all other collectors not authorized," Ibid., p. 10.

149 "the greatest the world has so far known," Schuchert and LeVene, op. cit., p. 197.

150 "The cost involved," Ibid., p. 453.

152 the piece exaggerated, Ostrom and McIntosh, op. cit., pp. 17–18.

152 "It would be well to hasten operations," William Carlin, letter of April 1, 1878, Othniel Charles Marsh Papers, reel 3, frame 74.

152 "There is no doubt," Ostrom and McIntosh, op. cit., p. 18.

152 Williston himself had become, Shor, *Fossils and Flies,* p. 86.

153 "It is just merely," Ostrom and McIntosh, op. cit., pp. 21–22.

153 "The station consisted of," Kohl and McIntosh, op. cit., p. 84.

153 "the miners entered the train," Ibid., p. 91.
154 "The trains constantly passing," Ibid., p. 99.
154 "It was a lovely morning," Ibid., p. 104.
154 "Professor Marsh, full of excitement," Ibid., p. 106.
154 "into a frenzy of patrol activity," Lanham, op. cit., p. 180.
154 "a tall, rather interesting-looking man," Kohl and McIntosh, op. cit., pp. 130–31.
155 "I felt that I was going," Osborn, *Master Naturalist*, p. 259.
155 "Today I made a grand exploration," Idem.
155 "what he felt they wanted to hear," Davidson, op. cit., p. 26.
155 "The *Monstrum horrendum*," Ostrom and McIntosh, op. cit., p. 29.
156 "My work has been solitary," Kohl and McIntosh, op. cit., p. 136.
156 "Collecting at this season," Ostrom and McIntosh, op. cit., pp. 32–34.
157 "They did not come to the quarey," Ibid., p. 38.
157 "This country is run over," Ibid., p. 39.
158 "a man working for you," Ibid., p. 46.
159 "well documented," Edwin Colbert, *Great Dinosaur Hunters*, p. 90.
160 "Marsh and Cope were almost," Ibid., p. 92.
161 "This is mentioned," Osborn, *Master Naturalist*, p. 258.
161 A book of memorial addresses, "Addresses in Memory of E. D. Cope," delivered at the American Philosophical Society, November 12, 1897.
162 "probably his greatest work," Schuchert and LeVene, op. cit., p. 377.
162 "more than a score of years," Lanham, op. cit., p. 183.
162 "In a total of three-and-a-half days," Ibid., p. 184.
163 "suddenly dinosaurs became famous worldwide," *Dinosaurs: The Monsters Emerge*, PBS, 1992.
163 "Ringmaster Marsh," Schuchert and LeVene, op. cit., p. 198.
163 accurately reflected contemporary public perceptions. Adrian Desmond wrote in *The Hot-Blooded Dinosaurs*, p. 116, that "the great extinct lizards had fired the public's imagination and new discoveries were eagerly reported in the popular press" during the 1840s, but he noted in *Archetypes and Ancestors*, p. 14, that the British public had lost interest in dinosaurs by the 1870s.
164 "expressing his regret," "A Field Day Among Geologists," *New York Tribune*, April 28, 1877, p. 5.
164 "Its remains were found," "Treasures of Science," *New York Tribune*, August 11, 1877.
164 "a solitary ichthyosaurus," *New York Times*, September 7, 1877.
164 "pterodactyl-plesiosaur," *New York Tribune*, October 10, 1877.

164 "It is not every one," *New York Times,* October 20, 1877, p. 4.

165 John S. Newberry, professor of geology, "Colorado's Antique," *New York Tribune,* December 8, 1877.

165 calling for public dissection, *New York Times,* December 8, 1877, p. 5.

165 "Professors Taylor and Paige," "A Petrified Man," *New York Herald,* December 8, 1877, p. 8.

165 "Professor Marsh was again called upon," G. A. Stockwell, "The Cardiff Giant, and Other Frauds," *Popular Science Monthly,* vol. 13, p. 202.

166 "There are probably some," "Mysteries of the Deep," *New York Herald,* October 9, 1877.

166 Marsh had paid $1000 for a pterodactyl, *New York Herald,* July 8, 1877.

166 "blood stained the snow," O'Connor, op. cit., p. 139.

11. An Inside Job

167 "This is almost part of," Osborn, *Master Naturalist,* p. 235.

167 "through his indifference," Cope, "The Batrachia of the Permian Period of N.A.," p. 35.

168 "Cope began his classic series of papers," Osborn, *Master Naturalist,* p. 233.

168 Cope's Permian fossils, Desmond, *Huxley,* p. 501.

168 "Marsh distinguished himself," *New York Herald,* January 12, 1890.

168 "There is no truth," *New York Herald,* January 19, 1890.

168 "hitherto no Permian vertebrates," Marsh, "Notice of New Fossil Reptiles," p. 409.

169 "After I read a paper," *New York Herald,* January 20, 1890.

169 "His work of identification," Alfred S. Romer, "Cope Versus Marsh," p. 204.

170 "was more than an illustrious scientist," Stegner, op. cit., p. 232.

170 "It was a tight inside job," Ibid., p. 234.

171 "wonderfully like something I have read," Ibid., p. 235.

171 "simply the political manager of his expedition," Lanham, op. cit., p. 226.

171 "I have often heard his ignorance," Ibid., p. 227.

171 "splendid appropriations," Foster, op. cit., p. 314.

172 "had everything to interest," Henry Adams, *The Education of Henry Adams,* p. 311.

172 "understood the congressman," Stegner, op. cit., p. 249.

174 "Its most striking room," Schuchert and LeVene, Stegner, op. cit., pp. 342–48.

175 "Newspaper men were particularly guarded against," Erwin H. Barbour, "Notes on the Paleontological Laboratory of the United States Geological Survey Under Professor Marsh," *American Naturalist*, April 1890, p. 397.

175 "Not only was he eminent," Stegner, op. cit., p. 232.

176 "The summer of 1879," Osborn, *Master Naturalist*, pp. 269–72.

12. The Slippery Slope

178 His capital remained intact, Davidson, op. cit., p. 147.

179 "Whatever prizes he wanted," Adams, op. cit., p. 313.

179 "ideal, and his personal fitness," Ibid., p. 346.

179 "eschewing liquor and tobacco," Persifor Frazer, "The Life and Letters of Edward Drinker Cope," *American Geologist*, vol. 26, no. 2, p. 126.

180 "I think sooner or later," Osborn, *Master Naturalist*, p. 280.

180 "Huge amounts of gold," Lanham, op. cit., p. 235.

180 "A cold wind," Osborn, *Master Naturalist*, p. 295.

180 "The Philadelphians," Ibid., p. 281.

181 "Sage brush or nothing," Ibid., p. 285.

181 "The weather was," Ibid., p. 302.

181 "dragged his feet," Stegner, op. cit., p. 285.

181 "I have been here," Osborn, *Master Naturalist*, p. 304.

182 "Can't we scotch Powell?" *New York Herald*, January 12, 1890.

182 "was not interested in puddling old blood," Stegner, op. cit., p. 285.

182 "With childish confidence," Frazer, op. cit., p. 127.

182 "When I came here," Osborn, *Master Naturalist*, pp. 288–89.

183 "I thought it rather rough," Ibid., p. 290.

183 "by a judicious use of aconite," Ibid., p. 332.

183 "The only difficulty," Ibid., pp. 306–7.

183 "This kind of work," Ibid., p. 308.

184 "I am writing by a fire," Ibid., p. 310.

184 "many tales of ladies," Davidson, op. cit., pp. 108–9.

184 "with huge families," Osborn, *Master Naturalist*, p. 290.

184 "I am very sorry," Ibid., pp. 313–14.

184 "With mines it is either," Ibid., p. 291.
185 "Altogether it looks," Idem.
185 interspersed with evening conversation, Ibid., p. 290.
185 "I have the prospect," Ibid., p. 292.
185 "I have so far escaped malaria," Ibid., p. 342.
185 "My health is very good," Ibid., p. 293.
186 "I cannot employ these people," Ibid., p. 379.
186 "Absence from home," Ibid., p. 293.
186 "I found that certain things," Idem.
187 "I had my interview," Ibid., pp. 362–63.
188 "I suspect that my chances," Ibid., p. 363.
188 "So far as words go," Ibid., p. 364.
188 *"My bill is thrown out,"* Ibid., p. 366.
188 "I received as many compliments," Ibid., p. 367.
189 He didn't get it, Idem.
189 "Today I had," Ibid., p. 368.
189 "Powell says he cannot," Ibid., p. 369.
189 "My entire future," Ibid., p. 320.
190 "As this is the second time," Ibid., p. 326.
190 "except one to be paid for," Ibid., p. 327.
190 "endless trouble," Ibid., p. 393.
190 "I had many queer experiences," Ibid., p. 585.
190 "Several persons asked," Ibid., p. 325.
190 "The financial situation," Ibid., p. 327.
191 "I do not know where," Ibid., p. 378.
191 "The ethics of business," Ibid., p. 375.

13. Behind the Arras

192 "Few men have succeeded," Frazer, op. cit., p. 68.
192 "Inferior men," Osborn, *Master Naturalist,* p. 362.
193 "On the following morning," Ibid., p. 584.
193 "humorous Celtic attitude," Ibid., p. 408.
193 "particularly relished," Ibid., p. 585.
193 "met honest opposition," E. C. Case, "Cope—The Man," *Copeia,* no. 2, p. 62.
193 "He hated opponents," Idem.

193 "One day," Osborn, *Master Naturalist,* pp. 583–85.
193 "The geological survey," Stegner, op. cit., p. 284.
194 "plugged all the time," Schuchert and LeVene, op. cit., p. 9.
194 "a collection of wives," Ibid., pp. 353–54.
195 "not normally or properly constituted," Davidson, op. cit., p. 84.
195 "My belief is," Schuchert and LeVene, op. cit., p. 345.
195 "In due time," Ibid., p. 308.
195 "I had begged permission," Shor, *Fossils and Flies,* p. 90.
196 "Not only does he avoid," Barbour, op. cit., p. 396.
196 "Harger was Marsh's eyes," Lanham, op. cit., p. 246.
197 "To my personal knowledge," Samuel Williston, "Oscar Harger," *American Naturalist,* vol. 21, no. 12, p. 1133.
197 "drive and determination," Schuchert and LeVene, op. cit., p. 310.
197 "the Prussian laboratories," Lanham, op. cit., p. 247.
198 "that Marsh recognized," Schuchert and LeVene, op. cit., p. 305.
198 "He was angry," Shor, *Fossils and Flies,* p. 110.
199 "in part true," *New York Herald,* January 19, 1890.
199 "It was a Saturday afternoon," Idem.
200 "we were directed," Barbour, op. cit., p. 396.
200 "By Professor Silliman's invitation," *New York Herald,* January 20, 1890. Cope's taunt about the specimen that had been restored with plaster "in such a way as to deceive everybody" referred to allegations that Marsh artificially made his fossils seem more complete than they were. Erwin Barbour devoted much of his April 1890 *American Naturalist* article to these, implying that Marsh's fossils were so "restored" as to be nearly worthless. "Can the people see Government specimens?"
he asked. "No, they cannot! And in all justice to the present management, perhaps there is no reason, as he claims, why they should."
200 "On what meat does this our Caesar feed," Stegner, op. cit., p. 285.
200 "I had a letter," Osborn, *Master Naturalist,* p. 379.
201 "As to the Washington position," Ibid., p. 361.
201 "I presume you are aware," Stegner, op. cit., p. 289.
202 "place themselves and him," Osborn, *Master Naturalist,* p. 380.
202 "It should, if possible," Ibid., p. 382.
203 "It was somewhat unfortunate," Stegner, op. cit., p. 292.
203 "In a few days," Osborn, *Master Naturalist,* p. 373.

203 National Academy of Sciences meeting, Schuchert and LeVene, op. cit., p. 311.

203 "It is now I think," Osborn, *Master Naturalist,* p. 388.

204 "I have been at work," Ibid., p. 322.

204 "newspaper man of New York," Ibid., pp. 381–82.

204 "the first, last and almost the only," William Hosea Ballou, "Some Great American Scientists: VII. Edward Drinker Cope," *The Chatauquan,* vol. 50, p. 103.

204 "some of his work," Shor, *The Fossil Feud,* p. 60.

204 "perhaps the oldest river in existence," William Hosea Ballou, "Proposal for an Adirondack National Park," *American Naturalist,* vol. 19, no. 6, p. 578.

204 " 'Ancestor,' by the way," Ballou, "Some Great American Scientists," p. 117.

205 "pseudo-science," Shor, *The Fossil Feud,* p. 60.

205 "said the sea serpent," Osborn, *Master Naturalist,* p. 518.

205 "Can we pass him by," Ballou, "Some Great American Scientists," p. 101.

205 "I rushed over to see him," Ibid., pp. 104–5.

206 "I don't know where," Scott, op. cit., p. 231.

206 "I have endless trouble," Osborn, *Master Naturalist,* p. 393.

206 "Marsh had been in so long," Plate, op. cit., p. 230.

206 "I see the National Academy," Osborn, *Master Naturalist,* p. 391.

207 "I hear that Marsh," Ibid., p. 329.

207 Although no documented source. R. W. Howard's *The Dawnseekers,* p. 253, says that Joseph Leidy urged the University of Pennsylvania Board of Trustees to hire Cope, but gives no reference.

207 "he was extremely unpopular," Scott, op. cit., p. 232.

14. Cope Strikes

209 "O. C. Marsh has had more," Davidson, op. cit., p. 91.

209 "Secretary Noble," Osborn, *Master Naturalist,* p. 402.

210 "the move was pure malice," Lanham, op. cit., p. 249.

210 "persuaded Cope," Osborn, *Master Naturalist,* p. 402.

210 "I don't think writing private letters," Ibid., p. 411.

210 "fifty-two-column," Ballou, "Some Great American Scientists," p. 104.

210 "a young newspaper man," Osborn, *Master Naturalist,* p. 402.

210 "a journalist on the staff," Romer, op. cit., p. 206.

210 "*Herald* reporter," Stegner, op. cit., p. 325.

211 "After removing the unconfirmed," Shor, *The Fossil Feud,* pp. 55–61.

212 "I had heard of him," Scott, op. cit., pp. 218–19.

212 that Ballou had peddled, "An Old Grievance Aired," *New York Times,* January 13, 1890.

213 "two-way stream of memos and cables," O'Connor, op. cit., p. 161.

214 "Some weeks later," Scott, op. cit., p. 219. Osborn wrote that Ballou, "actuated only by an unbridled sense of publicity," had submitted the article to Powell and Marsh "to guard himself against libel and lay in a fresh stock of sensational material for the *Herald*" (*Master Naturalist,* p. 403). But Marsh's angry letter to Scott referred to the *Herald,* not to Ballou. It would have been the *Herald* zombies who took those clever and slippery moves, not the slippery but foolish Ballou.

214 "involved in a blackmailing case," Scott, op. cit., p. 219.

214 "Marsh has made a dead set," Osborn, *Master Naturalist,* p. 409.

215 "The *Herald,* getting wind," Scott, op. cit., p. 219.

215 he cabled from Ceylon, O'Connor, op. cit., p. 162.

216 "What he dreaded," Ibid., p.7.

222 "restraint and decorum," Stegner, op. cit., p. 327.

222 "newspapers throughout the country," Ibid., p. 326.

224 "simply a fancy picture," *New York Herald,* November 9, 1874, p. 3.

225 "I want you fellows to remember," Seitz, op. cit., p. 222.

15. The *Herald* Steams Ahead

226 "I hope a clean job may be made of it," Osborn, *Master Naturalist,* p. 410.

226 "All the old charges," Stegner, op. cit., pp. 324–25.

226 "thoroughly investigated," "An Old Grievance Aired," *New York Times,* January 13, 1890.

227 "I do not say," Idem.

227 "to Professor Cope's consuming restlessness," "Scientists at War," *Philadelphia Inquirer,* January 13, 1890, p. 2.

227 "Poor old Leidy," Osborn, *Master Naturalist,* p. 411.

227 "When hot news was involved," Kluger, op. cit., p. 143.

232 "There is one favor," Osborn, *Master Naturalist,* p. 410.

233 "had combined to enrich," "Prof. Marsh Denies It," *New York Times,* January 14, 1890, p. 6.

233 "Professor Marsh of Yale," "The Attack on Professor Marsh," *New Haven Palladium, New York Tribune,* January 14, 1890, p. 5.

236 "Apparently deeply affected," Adler, op. cit., p. 61.

16. Marsh Strikes Back

238 "Ballou misrepresented me," Osborn, *Master Naturalist,* p. 411.

238 "When asked about rumors," "Cope May Be Removed," *Philadelphia Inquirer,* January 14, 1890, p. 2.

239 "marked by no such restraint," Stegner, op. cit., p. 327.

239 "full and delighted," Ibid., p. 324.

245 "The recklessness of assertion," Cope, "On Professor Marsh's Criticisms." p. I.

245 "Whereas Cope attacked," Osborn, *Master Naturalist,* p. 405.

246 "it must have made Cope," Gus Craven, letter of March 3, 1891, Othniel Charles Marsh Papers, reel 3, frame 986.

17. The *Herald* Steams Away

247 "Cope had shot off," Stegner, op. cit., pp. 327–28. Url Lanham and Robert Plate seem to have drawn on Stegner's version. Plate wrote in *The Dinosaur Hunters,* p. 247, that Cope "took a lofty tack with Ballou" in making the "recklessness" response. Lanham wrote in *The Bone Hunters,* p. 259, that Cope "tried to wind up the affair" with it. Both Lanham and Plate cited Marsh's "mule's tail" taunt as the last word in the newspaper confrontation. The *Herald*'s bone-war coverage has confused other writers in other ways. Douglas J. Preston's *Dinosaurs in the Attic,* pp. 78–79, attributed Powell's masterly indictment of Cope to Marsh.

248 "It is equally forceful," Osborn, *Master Naturalist,* p. 411.

249 "Whenever there is an important piece," O'Connor, op. cit., p. 168.

250 "Paleozoic Poetry," Shor, *The Fossil Feud,* p. 334.

252 "one of the worst rings," Ballou, "Some Great American Scientists," p. 104.

252 "discomfited and even discredited," Stegner, op. cit., p. 328.

252 "Marsh's able rejoinder," Plate, op. cit., p. 247.

252 "I astonished the committee," Andrew Carnegie, letter of May 12, 1890, Othniel Charles Marsh Papers, reel 3, frames 81–85.

253 "the Cope-Ballou article," Osborn, *Master Naturalist,* pp. 405–8.

253 "not likely to dislodge him," Scott, op. cit., pp. 220–21.

253 "A scandal of no ordinary," "War Among the Scientists," *Chicago Tribune,* January 16, 1890, p. 4.

254 "The press has taken," Cope, "Editorial," *American Naturalist,* February 1890, pp. 158–59.

254 "The modern Naboth," Ibid., p. 159.

254 "This huge steal," Frazer, op. cit., p. 113.

18. Symmetries and Ironies

255 "making an exhibition," Frazer, op. cit., p. 114.

255 "A smear never quite," Stegner, op. cit., p. 328.

256 "Somewhere in the room," Ibid., p. 329.

256 "the major defeat," Ibid., p. 337.

257 "That was more than ominous," Ibid., p. 339.

257 "Again, as in the newspaper," Ibid., p. 340.

257 called the defeat "unexpected," "A Defeat for the Geological Survey," *New York Tribune,* July 15, 1892, p. 5.

257 "the usual attack and defense," "Proceedings in the Senate and House of Representatives," *New York Herald,* July 15, 1892, p. 4.

257 "the biggest cut," *New York Herald,* July 18, 1892, p. 10.

257 "That brought the house down," Stegner, op. cit., p. 340.

258 "Appropriation cut off," Schuchert and LeVene, op. cit., p. 320.

258 "I do not think," Ibid., p. 321.

258 "Marsh's world," Ibid., pp. 322–23.

260 "Who cares about Carson Valley," *New York Herald,* March 14, 1897, section 6, p. 3. *Mylodon* and *Megatherium* were giant sloths and looked very little like the "giant saurian creatures" in the *Herald* engraving. They were herbivorous, and probably not very aggressive.

260 "the old-timers," Shor, *The Fossil Feud,* p. 218.

261 "not as a scientist or partisan," Schuchert and LeVene, op. cit., p. 348.

261 His friend, the English paleontologist, Ibid., p. 342.

261 "I feel cheerful," Osborn, *Master Naturalist,* pp. 416–17.

261 "I never saw," Ibid., pp. 417–19.
262 "very poor success," Ibid., pp. 419–20.
262 "I find," Ibid., pp. 421–23.
262 "The trip involved," Ibid., p. 429.
262 "So between hands," Ibid., p. 438.
263 "Many were the narrow escapes," Ibid., p. 435.
263 "played across the sky," Ibid., p. 432.
263 "As it grew late," Idem.
263 "The lightning did not avenge," Idem.
264 "The Indians believe," Ibid., p. 442.
264 "It is very possible," Frazer, op. cit., p. 114.
264 "As he left President Jesup's office," Osborn, *Master Naturalist,* pp. 446–47.
265 "No disappointment," Frazer, op. cit., p. 69.
265 "Although I spoke to nobody," Osborn, *Master Naturalist,* p. 457.
265 "left him," Davidson, op. cit., p. 108.
266 "led him to extraordinary lengths," Schuchert and LeVene, op. cit., p. 339.
267 "Recently, for example," Hellman, "Go on Investigators! Scrutinize!" p. 76.
267 Observing at an alumni dinner, *New Haven Register,* December 26, 1897.
267 "In extinct Mammals," Schuchert and LeVene, op. cit., p. 327.
267 "The dinosaurs seem," Ibid., p. 385.
268 "This recognition," Ibid., p. 326.

19. Death

269 "avoid horseback riding," Osborn, *Master Naturalist,* pp. 584–85.
269 "slight attacks," Davidson, op. cit., p. 155.
270 "The Dr. says," Osborn, *Master Naturalist,* p. 465.
270 "even severe pain," Ibid., p. 463.
270 "I am not particular," Ibid., p. 450.
270 "I am back in Washington," Ibid., p. 495.
270 "Cope was old and weary," Ibid., p. 586.
271 "stomach or chest trouble," Ibid., p. 465.
271 "For two weeks," Ibid., p. 467.

271 "aghast at hearing," Ibid., p. 462.
271 "Cope for the first time," Ibid., pp. 586–87.
271 "I did not suffer," Ibid., p. 467.
271 "Third day noon," Ibid., p. 468.
272 "a great naturalist," Osborn, "A Great Naturalist: Edward Drinker Cope," p. 10.
272 "the most animal" mind, Davidson, op. cit., p. 109.
273 "In his laboratory," Ballou, "Strange Creatures of the Past," p. 15.
273 (probably by a *Century Magazine* "layman editor"), Ibid., p. 22.
273 "Professor O. C. Marsh's prior reconstruction," Ibid., p. 18.
274 "I don't expect," Osborn, *Master Naturalist,* p. 564.
274 "He became very reticent," Ibid., p. 545.
274 "last relapse," Ibid., p. 462.
274 "bright and animated," Ibid., p. 587.
275 "He was confined," Ibid., p. 468.
275 "Not many days," Idem.
275 "I have not been able," Davidson, op. cit., p. 159.
275 "for the last time," Osborn, *Master Naturalist,* pp. 587–89.
276 "a peer of Huxley and Owen," J. S. Kingsley, *American Naturalist,* vol. 31, May 1897, pp. 414–19.
276 "views and convictions," Persifor Frazer, Ibid., p. 412.
276 "his published works large and small," *American Journal of Science,* May 1897.
277 "a full course of Medicine," *New York Times,* May 13, 1897, p. 9.
277 "Professor Cope took an active part," *New York Herald,* April 13, 1897, p. 9.
278 "I have had enough," Schuchert and LeVene, op. cit., pp. 325–26.
278 "had been ill during that year," Ibid., p. 330.
278 "When the *Century,*" Ballou, letter of February 11, 1898, Osborn Library, American Museum of Natural History.
278 "as I had grossly misrepresented," Idem.
279 "be simultaneously attacked," Ballou, letter of January 22, 1898, Osborn Library.
279 "I am wholly familiar," Ballou, letter of February 4, 1898, Osborn Library.
279 "I think he will be compelled," Ballou, letter of February 11, 1898, Osborn Library. Ballou's letterheads are instructive. His June 19, 1897, letter was from "The Hoffman House"; his January 22, 1898, letter was

from "The New York and Westchester Water Company . . . Removed to the Morris Building"; his February 4 and 11 letters were from "The New York Dispatch," of which he signed himself "Editor."

279 "I have been much interested," Osborn, letter of February 13, 1898, Osborn Library.

280 "Scientists who have had," "Petrified Remains of a Race of Giants," *New York Herald,* January 2, 1898.

280 "I of course regretted," Marsh, "My First Visit to the Rocky Mountains," *Fossil Hunting in the Rocky Mountains,* Othniel Charles Marsh Papers, reel 26, frame 249.

281 NEW YORK'S NEWEST, *New York World,* December 26, 1898, p. 10. A certain respect for facts seems to have accompanied the *World*'s dinosaur sensationalism. Its headlines about 500-year-old, continually eating sauropods may be more or less scientifically sound.

282 "The last of February, 1899," Schuchert and LeVene, op. cit., pp. 330–31.

283 "leading men of science," Charles E. Beecher, "Othniel Charles Marsh," *American Journal of Science,* 4th ser., vol. 2, pp. 402–19.

283 "One smiles a little," Schuchert and LeVene, op. cit., p. 331.

283 "among the greatest investigators," *New York Times,* March 19, 1899, p. 6. The *Journal of the New Haven Historical Society,* vol. 19, no. 3, September 1970, p. 60, mentions Marsh's being approached by an unexpectedly urbane Indian during one of his expeditions—not a theology graduate, but a former schoolmate of another Yale professor.

284 "He had a controversy," *New York Herald,* March 19, 1899, p. 7.

284 His paper's circulation, Seitz, op cit., p. 371.

285 ("Say Bennett!" . . .), Ibid., p. 239.

285 "Is the *Herald* for sale?" O'Connor, op. cit., p. 268.

285 a "Scotch miser," Seitz, op. cit., p. 376.

285 "a thoroughly domesticated man," O'Connor, op. cit., p. 303.

285 to which he succumbed a few days later, Ibid., p. 315.

286 "There was a deeper sagacity," Ibid., p. 317.

20. The Skeleton Drummer

287 "At the close," Osborn, *Master Naturalist,* pp. 589–90.

287 "after my funeral," Ibid., p. 590.

288 "There are three hundred men," *New York Herald,* September 4, 1898, section 5, p. 6. The *Herald* also revealed that many of the "three hundred men" had secretly willed their brains to the Anthropometric Society, so it is possible that Marsh's was among them. Edward Spitzka's 1903 study did not mention Marsh's brain.

289 "The specimen is remarkable," Caroline Werkley, "Professor Cope, Not Alive but Well," *Smithsonian,* vol 6, no. 5, p. 73.

290 "Professor Eiseley removed the bones," Ibid., pp. 74–75.

290 "The skull lay tilted," Loren Eiseley, *The Immense Journey,* p. 3.

291 "sometime in the mid-1970s," Gale E. Christiansen, *Fox at the Wood's Edge,* p. 436.

291 "The skeletal bones," Alan Mann, letter of August 5, 1997.

291 "A very careful comparison," Alan Mann, letter of January 21, 1998.

292 "expressed to some of his colleagues," Werkley, op. cit., p. 75.

292 "Hahn took the bones," Christiansen, op. cit., pp. 436–37.

292 "After his death," Davidson, op. cit., p. 161.

293 "shook his head in disbelief," Louie Psihoyos and John Knoebber, *Hunting Dinosaurs,* pp. 20–21.

293 "the skull would remain at the Academy," Alan Mann, letter of August 5, 1997.

293 "weird unaccountable blue glow," Psihoyos and Knoebber, op. cit., p. 27.

293 "Everyone wanted to meet," Ibid., p. 22.

294 "By actually becoming," Ibid., p. 20.

294 "had done a search of the current literature," Ibid., p. 27.

294 Linnaeus *had* described, Earl E. Spamer, "Know thyself: Responsible Science and the lectotype of *Homo sapiens* Linnaeus, 1758," *Proceedings of the Academy of Natural Sciences,* vol. 149.

294 "It's just something," Ted Daeschler, telephone conversation of July 1997.

294 "This is something," Alan Mann, letter of September 23, 1997.

295 "stated flatly," Davidson, op. cit., p. 211.

295 "confident" that science would prove, *Dinosaurs: The Monsters Emerge,* PBS, 1992.

296 "a fervent racist, sexist," Mark Bowden, "On the Trail of a Wayward Skull," *Philadelphia Inquirer,* October 6, 1994, A–1.

296 "coincides not only," Cope, *The Origin of the Fittest*, p. 288.
296 "a little on the short side," Bowden, op. cit.
296 "Professor Edward Cope," Psihoyos and Knoebber, op. cit., p. 29.
296 "the museum had asked the FBI," Alan Mann, letter of August 5, 1997.

Epilogue: Squabblers on a Raft

298 "It has been well said," Marsh, "My First Visit to the Rocky Mountains," pp. 1–2, Othniel Charles Marsh papers, reel 26, frames 246–47.
299 Cope was said to have stolen a whale skeleton, Romer, op. cit., p. 202.
300 "Joseph Leidy," Thomas Beer, *The Mauve Decade*, p. 172.
300 "the growing cult of bone digging," Bowden, op. cit.
301 an example of racist evolutionism, Stephen Jay Gould, *The Mismeasure of Man*, p. 116. Gould quotes Cope: "The Greek nose, with its elevated bridge, corresponds not only with aesthetic beauty, but with developmental perfection."
301 "maintained a drumfire of criticism," E. O. Wilson, *Naturalist*, p. 341.
301 "the most peculiar," Stephen Jay Gould, "Cardboard Darwinism," p. 28.
301 "a renaming," E. O. Wilson, *The Diversity of Life*, p. 89.
301 Gould condemned the assault as "infantile," Wilson, *Naturalist*, p. 350.
302 "I became determined," Wilson, *Naturalist*, p. xii.
303 "When I focused," Ibid., p. 86.
303 "In my heart I will be," Ibid., p. 163.
303 "I was the naturalist," Ibid., p. 345.
303 "At my core," Ibid., p. 25.
303 (he called John Ruskin "a monumental ass" . . .), Davidson, op. cit., p. 121.
303 "Science is modern civilization's," Wilson, *Naturalist*, p. 27.
303 "The fourth floor," Gould, "Evolution by Walking," p. 253.
304 "Great scientists," Gould, "Natural Selection and the Human Brain," p. 47.
304 "When I think of paleontology," Gould, *The Panda's Thumb*, p. 14.
304 "Most of all," Gould, "The Telltale Wishbone," p. 267.
305 "a collection of eclectic amateurs," Gould, *The Mismeasure of Man*, p. 31.
305 as though the thirteen colonies had never existed, Gould, "Opus 100," p. 183.

305 "Nothing arouses," Gould, "Biological Potentiality versus Biological Determinism," p. 251.

305 "I was unprepared," Wilson, *Naturalist,* p. 339.

305 "one theme of transcendent," Gould, *Eight Little Piggies,* p. 13. In one essay, "The Golden Rule," Gould wrote, "The conservation movement was born, in large part, as an elitist attempt by wealthy social leaders to preserve wilderness as a domain for patrician leisure and contemplation (against the image, so to speak, of poor immigrants traipsing through the woods with their Sunday picnic baskets)." Gould probably had in mind the wealthy Henry Fairfield Osborn, who was instrumental in saving old growth coast redwoods, but he failed to note that early conservation's two major figureheads, Muir and Audubon, were poor immigrants.

306 "canonical stories of the extinction saga," Gould, "Losing a Limpet," p. 54.

307 "The so-called wise-use," Michael Satchell, "Dinosaur Bone Wars," *U.S. News and World Report,* August 26, 1996, pp. 43–45.

308 "transformed paleontology," Björn Kurtén, *The Age of Dinosaurs,* p. 24.

308 "When the century," Wilson, *Naturalist,* p. xi.

308 As perennial polls show. In a 1991 Gallup Poll, 40 percent of Americans polled believed that all life was created less than ten thousand years ago and another 47 percent believed that, although life may be millions of years old, it was created in more or less its present form. Nine percent believed life evolved gradually from primitive forms, and 4 percent were unsure.

Bibliography

Adams, Henry. *The Education of Henry Adams.* New York: Modern Library, 1931.

Adler, Kraig (editor). *Contributions to the History of Herpetology.* Society for the Study of Amphibians and Reptiles, 1989.

Bakker, Robert T. *The Dinosaur Heresies.* New York: William Morrow, 1986.

Ballou, William Hosea. "Proposal for an Adirondack National Park." *The American Naturalist,* vol. 19, no. 6 (June 1885).

———"Some Great American Scientists: VII. Edward Drinker Cope." *The Chautauquan,* vol. 50 (March–May 1908).

———. "Strange Creatures of the Past." *The Century Illustrated Magazine,* vol. 55, November 1897, pp. 15–23.

Barbour, Erwin H. "Notes on the Paleontological Laboratory of the United States Geological Survey Under Professor Marsh." *The American Naturalist,* April 1890.

Bartram, William. *Travels.* New York: Dover, 1955.

Beecher, Charles E. "Othniel Charles Marsh." *American Journal of Science,* fourth series, vol. 2, June 1899, 402–19.

Beer, Thomas. *The Mauve Decade: American Life at the End of the Nineteenth Century.* New York: Alfred A. Knopf, 1926.

Boorstin, Daniel J. *The Americans: The Colonial Experience.* New York: Random House, 1958.

———. *The Americans: The Democratic Experience.* New York: Random House, 1973.

———. *The Americans: The National Experience.* New York: Random House, 1965.
Bowden, Mark. "On the Trail of a Wayward Skull." *Philadelphia Inquirer,* October 6, 1994, A–1.
Bowler, Peter J. *Life's Splendid Drama: Evolutionary Biology and the Reconstruction of Life's Ancestry, 1860–1940.* Chicago: University of Chicago Press, 1996.
Case, E. C. "Cope—The Man." *Copeia,* no. 2 (July 28, 1940) pp. 61–65.
Christiansen, Gale E. *Fox at the Wood's Edge.* New York: Henry Holt, 1990.
Colbert, Edwin H. *The Great Dinosaur Hunters and Their Discoveries.* Mineola, NY: Dover Publications, 1984.
Cook, James H. *Fifty Years on the Old Frontier.* Norman: University of Oklahoma Press, 1954.
Cope, Edward Drinker. "The Batrachia of the Permian Period of North America." *The American Naturalist,* vol. 18, no. 1 (Jan. 1884) p. 26.
———. "A Contribution to the Zoology of Montana." *The American Naturalist,* vol. 13 (June 1879).
———. "The Creodonta." *The American Naturalist,* vol. 18, no. 3 (1884) pp. 255, 344, 478.
———. "Descriptions of Extinct Batrachia and Reptilia from the Permian Formation of Texas." *Proceedings of the American Philosophical Society,* vol. 17 (1878) pp. 38–58.
———. "Fossil Reptiles of New Jersey." *The American Naturalist,* vol. 1, no. 1 (March 1867) pp. 23–30.
———. "The Gigantic Mammals of the Genus *Eobasileus.*" *The American Naturalist,* vol. 7, no. 3 (March 1873) pp. 157–60.
———. *On the Method of Creation of Organic Types* (pamphlet). Philadelphia: M'Calla and Stavely, 1871.
———. "On Prof. Marsh's Criticisms." *The American Naturalist,* vol. 7, no. 7 (July 1873) Appendix: pp. I–II.
———. "On Some of Professor Marsh's Criticisms." *The American Naturalist,* vol. 7, no. 5 (May 1873) pp. 290–99.
———. "On the Vertebrata of the Dakota Epoch of Colorado." *Proceedings of the American Philosophical Society,* vol. 16, pp. 233–47.
———. *The Origin of the Fittest.* New York: Macmillan, 1887.
———. "Synopsis of the Vertebrate Fauna of the Puerco Series," *Transactions of the American Philosophical Society,* vol. 16, 1888.
———. "The Tertiary Marsupalia." *The American Naturalist,* vol. 18, no. 7 (July 1884) pp. 687–97.

———. *The Vertebrata of the Tertiary Formations of the West: Book One.* Washington, DC: Government Printing Office, 1883.
———. "The Vertebrate Fauna of the Puerco Epoch." *The American Naturalist,* vol. 22, no. 254 (February 1888) pp. 161–63.
Crockett, Albert Stevens. *When James Gordon Bennett Was Caliph of Bagdad.* New York: Funk and Wagnalls, 1926.
Davidson, Jane Pierce. *The Bone Sharp: The Life of Edward Drinker Cope.* Philadelphia: Academy of Natural Sciences of Philadelphia, 1997.
Desmond, Adrian. *The Hot-Blooded Dinosaurs.* London: Blond and Briggs, 1975
———. *Huxley: From Devil's Disciple to Evolution's High Priest.* Reading, MA: Addison-Wesley, 1997.
Desmond, Adrian and James Moore. *Darwin: The Life of a Tormented Evolutionist.* New York: Warner Books, 1991.
Eiseley, Loren. *Darwin's Century: Evolution and the Men Who Discovered It.* Garden City, NY: Doubleday and Company, 1958.
———. *The Immense Journey.* New York: Random House, 1957.
Elman, Robert. *First in the Field: America's Pioneering Naturalists.* New York: Mason Charter, 1977.
Foster, Mike. *Strange Genius: The Life of Ferdinand Vandeveer Hayden.* Niwot, CO: Roberts Rhinehart Publishers, 1994.
Frazer, Persifor. "The Life and Letters of Edward Cope." *American Geologist,* vol. 26. no. 2 (August 1900) pp. 67–129.
Gallagher, William B. *When Dinosaurs Roamed New Jersey.* New Brunswick, NJ: Rutgers University Press, 1997.
Galusha, Ted and John C. Blick. *Stratigraphy of the Santa Fe Group, New Mexico.* New York: Bulletin of the American Museum of Natural History, vol. 144, 1971.
Geiser, Samuel Wood. *Naturalists of the Frontier.* Dallas: Southern Methodist University Press, 1948.
Goetzmann, William H. *Exploration and Empire: The Explorer and the Scientist in the Winning of the American West.* Austin: Texas State Historical Society, 1993.
Gould, Stephen Jay. "Biological Potentiality vs. Biological Determinism." *Ever Since Darwin.* New York: W. W. Norton, 1977, pp. 251–60.
———. "Cardboard Darwinism." *An Urchin in the Storm.* New York: W. W. Norton, 1987, pp. 26–50.
———. "Dinomania." *Dinosaurs in a Haystack.* New York: Harmony Books, 1995, pp. 221–37.

———. "Double Trouble." *The Panda's Thumb*. New York: W. W. Norton, 1980, pp. 35–47.
———. "Evolution by Walking." *Dinosaurs in a Haystack*, pp. 248–60.
———. "Losing a Limpet." *Eight Little Piggies*. New York: W. W. Norton, 1993, pp. 52–63.
———. *The Mismeasure of Man*. New York: W. W. Norton, 1981.
———. "Natural Selection and the Human Brain: Darwin vs. Wallace." *The Panda's Thumb*, pp. 47–59.
———. "On Heroes and Fools in Science." *Ever Since Darwin*, pp. 201–7.
———. *Ontogeny and Phylogeny*. Cambridge: Harvard University Press, 1977.
———. "Opus 100." *The Flamingo's Smile*. New York: W. W. Norton, 1985, pp. 167–85.
———. "The Telltale Wishbone." *The Panda's Thumb*, pp. 267–78.
Hellman, Geoffrey. "Onward and Upward with Science: Go On, Investigators, Scrutinize!" *The New Yorker*, November 3, 1962, pp. 142–76.
Hellman, Hal. *Great Feuds in Science: Ten of the Liveliest Disputes Ever*. New York: John Wiley and Sons, 1998.
Herbst, Josephine. *New Green World: John Bartram and the Early Naturalists*. New York: Hastings House, 1954.
Howard, Robert West. *The Dawnseekers: The First History of American Paleontology*. New York: Harcourt Brace Jovanovich, 1975.
Huxley, Leonard. *The Life and Letters of Thomas H. Huxley*. New York: Appleton and Company, 1900.
Hyde, George. *Red Cloud's Folk*. Norman: University of Oklahoma Press, 1937.
Kastner, Joseph. *A Species of Eternity*. New York: Alfred A. Knopf, 1977.
Kingsley, J. S. "Edward Drinker Cope." *The American Naturalist*, vol. 31, May 1897, pp. 414–19.
Kluger, Richard. *The Paper: The Life and Death of the Herald Tribune*. New York: Alfred A. Knopf, 1986.
Kohl, Michael F. and John S. McIntosh (editors). *Discovering Dinosaurs in the Old West: Field Journals of Arthur Lakes*. Washington, DC: Smithsonian Institution Press, 1997.
Kurtén, Björn. *The Age of Dinosaurs*. New York: McGraw Hill World University Library, 1968.
———. *The Age of Mammals*. New York: Columbia University Press, 1971.
Lanham, Url. *The Bone Hunters*. New York: Columbia University Press, 1973.
Larson, Robert W. *Red Cloud: Warrior Statesman of the Lakota Sioux*. Norman: University of Oklahoma Press, 1997.
Leidy, Joseph. *The Extinct Mammalian Fauna of Dakota and Nebraska*.

Introduction by Ferdinand Hayden. Philadelphia: Academy of Sciences, 1869.

Lucas, Spencer G. with J. Keith Rigby, Jr., and Barry S. Kues (editors). *Advances in San Juan Basin Paleontology*. Albuquerque: University of New Mexico Press, 1981.

——— and Robert M. Schoch. "Cope, Marsh, and the Type Specimen of *Lystrosaurus 'frontosus.'*" *Discovery*, vol. 15, no. 2 (1980–1981) pp. 29–33.

Marsh, Othniel Charles. "The Fossil Mammals of the Order Dinocerata." *The American Naturalist*, vol. 7, no. 3 (March 1873) pp. 146–53.

———. "Introduction and Succession of Vertebrate Life in America." *American Journal of Science*, third Series, vol. 14, nos. 79–84, July–December 1877, p. 340.

———. "Notice of New Fossil Reptiles." *American Journal of Science*, vol. 15, no. 89 (May 1878) pp. 409–11.

———. "Notice of a New and Gigantic Dinosaur." *American Journal of Science*, vol. 14 (July 1877) pp. 87–88.

———. "Observations on the Metamorphosis of Siredon into Amblystoma." *American Journal of Science*, vol. 46, no. 138 (November 1868) pp. 364–74.

———. "Odontornithes, or Birds with Teeth." *The American Naturalist*, vol. 9, no. 12 (December 1875).

———. "On the Dates of Professor Cope's Recent Publications." *The American Naturalist*, vol. 7, no. 5 (May 1873) pp. 303–6.

———. "On the Genus *Tinoceras* and Its Allies." *The American Naturalist*, vol. 7, no. 4 (April 1873) pp. 217–18.

———. "Reply to Prof. Cope's Explanations." *The American Naturalist*, vol. 7, no. 6 (June 1873) pp. I–IX.

———. *A Statement of Affairs at Red Cloud Agency, Made to the President of the United States by O. C. Marsh* (pamphlet). Yale College, July 10, 1875.

Mattes, Merrill J. *Indians, Infants, and Infantry*. Denver: Old West Publishing Company, 1960.

McCarren, Mark J. *The Scientific Contributions of Othniel Charles Marsh: Birds, Bones, and Brontotheres*. New Haven: Peabody Museum Special Publication, no. 15, 1993.

McGinnis, H. J. *Carnegie's Dinosaurs*. Pittsburgh: Carnegie Museum, 1982.

Mitchell. W. J. T. *The Last Dinosaur Book: The Life and Times of a Cultural Icon*. Chicago: University of Chicago Press, 1998.

O'Connor, Richard. *The Scandalous Mr. Bennett*. Garden City, NY: Doubleday and Company, 1962.

Osborn, Henry Fairfield. *Cope: Master Naturalist.* Princeton, NJ: Princeton University Press, 1931.

———. "A Great Naturalist: Edward Drinker Cope." *The Century Illustrated Magazine,* vol. 55, November 1897.

———. *Impressions of Great Naturalists.* New York: Charles Scribner's Sons, 1924.

Ostrom, John H. and John S. McIntosh. *Marsh's Dinosaurs: The Collections from Como Bluff.* New Haven and London: Yale University Press, 1966.

Penick, James. "Professor Cope vs. Professor Marsh," *American Heritage,* vol. 22, no. 5, August 1971, pp. 5–13.

Plate, Robert. *The Dinosaur Hunters.* New York: David McKay and Company, 1964.

Psihoyos, Louie and John Knoebber. *Hunting Dinosaurs.* New York: Random House, 1994.

Rainger, Ronald. *An Agenda for Antiquity: Henry Fairfield Osborn and Vertebrate Paleontology at the American Museum of Natural History.* Tuscaloosa: University of Alabama Press, 1991.

Riffenburgh, Beau: *The Myth of the Explorer: The Press, Sensationalism, and Geographic Discovery.* Oxford: Oxford University Press, 1994.

Rogers, Katherine. *A Dinosaur Dynasty: The Sternberg Fossil Hunters.* Missoula, MT: Mountain Press Publishing Company, 1991.

Romer, Alfred S. "Cope Versus Marsh." *Systematic Zoology,* vol. 13, no. 4 (December 30, 1964) pp. 201–7.

Rudwick, Martin J. S. *The Meaning of Fossils: Episodes in the History of Palaeontology.* Chicago: University of Chicago Press, 1985.

Satchell, Michael. "Dinosaur Bone Wars." *U.S. News and World Report.* August 26, 1996, pp. 43–45.

Schuchert, Charles and Clara M. LeVene. *O. C. Marsh: Pioneer in Paleontology.* New Haven: Yale University Press, 1940.

Scott, William Berryman. *Some Memories of a Paleontologist.* Princeton, NJ: Princeton University Press, 1939.

Seitz, Don Carlos. *The James Gordon Bennetts: Father and Son, Proprietors of the New York Herald.* Indianapolis: Bobbs-Merrill, 1928.

Shor, Elizabeth Noble. *The Fossil Feud.* Hicksville, NY: Exposition Press, 1974.

———. *Fossils and Flies: The Life of a Compleat Scientist, Samuel Wendell Williston.* Norman: University of Oklahoma Press, 1971.

Simpson, George Gaylord. "Hayden, Cope, and the Eocene of New Mexico."

Proceedings of the Academy of Natural Sciences of Philadelphia, vol. 103 (1951) pp. 1–26.

Spalding, David. A. E. *Dinosaur Hunters: Eccentric Amateurs and Obsessed Professionals.* Toronto: Key Porter Books, 1993.

Spamer, Earle E. "Know Thyself: Responsible Science and the Lectotype of *Homo sapiens* Linnaeus, 1758." *Proceedings of the Academy of Natural Sciences of Philadelphia,* vol. 149.

Spitzka, Edward A. "A Study of the Brains of Six Eminent Scientists and Scholars Belonging to the American Anthropometric Society." *Transactions of the American Philosophical Society,* vol. 21, no. 4, 1907, pp. 175–308.

Stegner, Wallace. *Beyond the Hundredth Meridian: John Wesley Powell and the Second Opening of the West.* Boston: Houghton Mifflin Company, 1954.

Sternberg, Charles H. *The Life of a Fossil Hunter.* San Diego: Jensen Printing Company, 1931.

Stockwell, G. A. "The Cardiff Giant, and Other Frauds." *Popular Science Monthly,* vol. 13, no. 9, June 1878, pp. 197–203.

Utley, Robert Marshall. *The Lance and the Shield: The Life and Times of Sitting Bull.* New York: Henry Holt, 1993.

Warren, Gouverneur Kemble. "Exploration in the Dacota Country in the Year 1855." *Senate Executive Document No. 76, 34th Congress, 1st. Session.* Washington, DC: Government Printing Office.

Warren, Leonard. *Joseph Leidy: The Last Man Who Knew Everything.* New Haven: Yale University Press, 1998.

Webb, William Edward. *Buffalo Land: An Authentic Account.* Cincinnati and Chicago: E. Hannaford and Co., 1873.

Werkley, Caroline. "Professor Cope, Not Alive but Well." *Smithsonian,* vol. 6, no. 5, pp. 72–75.

Wheeler, Walter H. "The Uintatheres and the Cope-Marsh War." *Science,* vol. 131 (1960) pp. 1171–76.

White, Andrew D. "The Cardiff Giant: The True Story of a Remarkable Deception." *The Century Magazine,* vol. 64, no. 6, October 1902, pp. 948–55.

Wilford, John Noble. *The Riddle of the Dinosaur.* New York: Alfred A. Knopf, 1985.

Williamson, Thomas E. and Spencer G. Lucas. "Paleocene Vertebrate Paleontology of the San Juan Basin, New Mexico." *Vertebrate Paleontology in New Mexico.* Albuquerque, NM: Museum of Natural History and Science, bulletin 2, 1993.

Williston, Samuel. "Oscar Harger." *The American Naturalist*, vol. 21, no. 12, December 1887.

Wilson, Edward O. *Consilience: The Unity of Knowledge*. New York: Alfred A. Knopf, 1998.

———. *The Diversity of Life*. Cambridge: Harvard University Press, 1992.

———. *Naturalist*. Washington, DC: Island Press, 1994.

Index